The Cod Fishery of Isle Royale, 1713-58

B.A. Balcom

Studies in Archaeology
Architecture and History

National Historic Parks and Sites Branch
Parks Canada
Environment Canada
1984

Available in Canada through authorized bookstore agents and other bookstores, or by mail from the Canadian Government Publishing Centre, Supply and Services Canada, Hull, Quebec, Canada K1A 0S9.

En français ce numéro s'intitule **La pêche de la morue à l'île Royale, 1713-1758** (n⁰ de catalogue R61-2/9-15F). En vente au Canada par l'entremise de nos agents libraires agréés et autres librairies, ou par la poste au Centre d'édition du gouvernement du Canada, Approvisionnements et Services Canada, Hull, Québec, Canada K1A 0S9.

Price Canada: $5.25
Price other countries: $6.30
Price subject to change without notice.

Catalogue No.: R61-2/9-15E
ISBN: 0-660-11559-X
ISSN: 0821-1027

Published under the authority
of the Minister of the Environment,
Ottawa, 1984.

Editing, layout, and design: Barbara Patterson

Cover: Drying cod on rocks and large stones. From Duhamel Du Monceau, Traité Général des Pesches et Histoire des Poissons..., Part II, Section 1, Sailant & Nyon et Disant, Paris, 1772.

The opinions expressed in this report are those of the author and not necessarily those of Environment Canada.

Parks Canada publishes the results of its research in archaeology, architecture and history. A list of titles is available from Research Publications, Parks Canada, 1600 Liverpool Court, Ottawa, Ontario K1A 1G2.

THE COD FISHERY OF ISLE ROYALE, 1713-58

B.A. Balcom

Submitted for publication in 1979 by B.A. Balcom, Fortress of Louisbourg National Park.

INTRODUCTION

> Le commerce de l'Isle Royale peut se diviser en cinq branches principales dont la pesche est la premiere et la baze de toutes celles qui ont raport a la france, au Canada, aux Isles meridionnales, et enfin comme le Trone qui nourit toutes les autres.[1]

With these words, Jacques Prévost, *commissaire-ordonnateur* of Isle Royale, described the importance of the fishery to his colony. He made this evaluation in 1753 when the fishery's primacy in the colonial economy was weakening. The level of production was well below that of the late 1730s, and other branches of commerce, notably trade with the French West Indies and New England, were dramatically increasing. Despite a decrease in its relative importance the fishery was still acknowledged as the leading sector of the colony's economy.

In spite of this primacy, historical analysis of this industry has been limited mainly to its treatment as one aspect of a larger topic. Charles De La Morandière and Harold Innis dealt respectively with Isle Royale as part of a national and an international history of the North American cod fishery; A.H. Clark and J.S. MacLennan analysed the fishery with regard to the general history of Isle Royale; and Mary C. MacDougall Maude and Christopher Moore discussed it in relation to specific topics in the colony's history.[2] Even when Isle Royale's economy, which principally meant the fishery, was analysed in its own right, the viewpoint was not necessarily local. In separate articles, Innis and Clark reviewed the colony's economy with emphasis on the unwholesome effects of its trade association with New England.[3] Until the writings on Isle Royale's economic and social history increase in number and quality, the popular image of Louisbourg as just a battlesite of empires will remain unshaken.

Unfortunately the lack of historiography on the fishery is not confined solely to Isle Royale. In contrast with the attention given the production of other staple products in North America's development, such as furs, lumber, and wheat, historians have largely neglected the fishery's effect on the Atlantic seaboard. Indeed, only Newfoundland, whose dependence on the fishery was at least as great as that of Isle Royale, has a historiography of any depth on the fishery. Further-more, much of the historical writing on the northwestern Atlantic fishery is concerned more with its political ramifications than with its social or economic impact. The relatively large volume of literature on the North Atlantic fisheries disputes of the 19th and early 20th centuries between British North America and the United States is perhaps the best example of this trend. The consequence of this general neglect and strong political bias has been a general lack of studies utilizing the newer analytic techniques and methodologies of economic and social history.

There is reason to hope for a reversal of this longstanding trend as a new generation of historians enters this field. For example, David Alexander, in such works as *A New Newfoundland: The Traditional Economy to 1934,* has used economic models and quantitative analysis to reach new conclusions about that island's economy and particularly about its inshore fishery.[4] Similarly Anne Yentsch is employing demographic techniques of family reconstitution and aggregate analysis in her studies of seaboard communities on Cape Cod, Massachusetts.[5] Although it is perhaps early to judge the effect of these new historians, a growing body of analytic techniques and comparative case studies can only aid and inspire historians working on the fisheries of other areas.

The present shortage of comparative studies places some limitations on the study of Isle Royale's fishery. Because of the dispersed and international nature of fish producers and consumers, Isle Royale was both directly and indirectly influenced by fluctuations in other production areas and markets. Standard secondary works such as Innis and De La Morandière provide a general perspective but very little detail on local economies. With regard to supportive works on the Isle Royale economy, the situation is somewhat brighter. Although historians have often ignored the colony's economy, the recent work of Christopher Moore has helped to correct this imbalance.[6] Many aspects of Isle Royale's economy still require detailed investigation and such gaps restrict the fullest historical analysis of the fishery. The absence of a detailed price index, for example, means that trends in cost and profitability can only be analysed in rudimentary terms.

The abundance of primary documentation on Isle Royale offsets to some extent the lack of comparative secondary literature for a study of the fishery within the colonial context. Although much of the documentation is official in nature and primarily reflects administrative concerns, there are frequent references to the fishery. The heavy governmental regulation of trade and industry in a mercantilist society made the performance of Isle Royale's fishery a particular concern of local officials in their reports to France. Annual reports on the success of the fishery are available for some years. These provide, with some shortcomings, an important source of statistical information and frequently offer generalized explanations for fluctuations in production. Imperial and colonial legislation provides insights into the real and imagined ills of the industry, as well as into the concepts influencing governmental involvement. Censuses give cross sections of the people working in the fishery at given points in time but their utility is undermined by the transient nature of much of the fishery's labour force. In general, official documents fail to give information on the operations and activities of individuals in what was essentially a private industry.

In this latter regard, the judicial and notarial records of Isle Royale form an important information source on individual business transactions. Although incomplete in both their initial compilation and survival, the notarized documents include debts incurred for the purchase of supplies and equipment, labour engagements, sales and rentals of shore concessions, and contracts for the construction of boats and vessels. Such documentation provides information on common business practices in the fishery and does much to overcome the general absence of private ledgers and account books so useful in such analysis. The court records provide similar information on these practices, albeit from the biased view of law enforcement. Estate inventories are another valuable source as they give indications of the material culture of the fishery and also of the levels of wealth attained by the various groups engaged in it. Yet even here the incompleteness of these records makes it difficult to trace individual careers and gather quantifiable data.

There are highly detailed descriptions of the French cod fishery that provide a wealth of valuable information on its technical and operational aspects. Nicolas Denys wrote a description of fishery practices some 60 years before the founding of Louisbourg; Duhamel Du Monceau published a similar work in 1769, 11 years after the town's second and final capture by the English.[7] By balancing these two accounts during a period of little technological change, a fairly precise picture of the fishery methodology utilized in Isle Royale can be derived. Additional information is found in several memoirs on the fishery included in the official documentation.[8] They were written by both administrators and private citizens who were often seeking redress during times of crisis in the industry. Although written with particular biases in mind, these memoirs offer comprehensive views of the fishery and serve as important checks on the more general sources.

Most of the available sources deal with the Isle Royale fishery from the colonial viewpoint. This study naturally reflects that perspective in its analysis of the fishery's economic importance, its methodology, the personnel involved, and its impact on society. Unfortunately, little can be said in a detailed way about the operating costs and profitability of individual operations.

Working as a staff historian at the Fortress of Louisbourg National Historic Park gives special opportunities to, and places special demands on, an individual in his preparation of an historical report. The most important of those demands is undoubtedly that the report be of practical use to the Park's reconstruction or animation programme. That is, the final report must include "applied" as well as "pure" research. It is hoped that a sufficient body of material is covered in this study to meet that goal, and that the reader will thereby gain an appreciation of the history and practices of the cod fishery of Isle Royale.

FISHING AND THE ECONOMY
OF ISLE ROYALE

Perhaps more than most places, the history of the colony of Isle Royale has been shaped by military and political factors rather than by economic ones. As the former factors have figured so prominently in the colony's historiography, they require only a brief mention here. The colony was founded in 1713 to offset French territorial losses in Acadia and Newfoundland by the Treaty of Utrecht and comprised the islands in the Gulf of St. Lawrence — principally Isle Royale (Cape Breton), Isle St. Jean (Prince Edward Island), and Isles de la Magdaline (Magdalen Islands). As both the colony and its principal island share the name of Isle Royale, Isle Royale will be used in this report to describe the colony unless the island is specified. Barely 30 years after its founding the colony was occupied by the English — its capital besieged and captured, its fishing outposts burned, and its population deported. After an English occupation of 4 years, the French retook possession of the colony in 1749. The second French period was considerably shorter than the first. In 1758 the English captured Louisbourg for the second and final time. The French lost final possession in 1763, by which time Louisbourg's fortifications had already been demolished.

French appreciation of Isle Royale's economic and strategic potential received written expression during the early years of the 18th century. Although the 1706 colonization proposal by Antoine-Denis Raudot, intendant of New France, is the one most frequently cited by historians,[1] it was only one of a number of such memoirs written at that time by Frenchmen interested in the island's development.[2] In general, these memoirs stressed the economic potential of the island's fisheries and its interior resources such as timber and coal. Isle Royale was also foreseen as an entrepôt within the French imperial system. Raudot wisely foresaw that for the future colony to succeed in this role, its trade would have to be international and not just imperial in scope. Finally, these memorialists viewed a French establishment on Isle Royale as a protection for French fishing efforts, a containment to English expansion, and an operational base in time of war for French naval ships and privateers acting against English shipping and coastal colonies. These strategic considerations have had an unfortunate historiographical legacy in which Isle Royale, and in particular Louisbourg, features as an expensive but ineffective "Guardian of the Gulf of St. Lawrence."[3]

Modifications occurred as the projections contained in these early memoirs materialized. Three main areas of economic endeavour developed but the colony's land resources, both extractive industries and agriculture, failed to achieve more than marginal significance in the colonial economy. In their place, government expenditure on Louisbourg's fortifications and other public spending on the colonial administration and on a relatively large garrison assumed major proportions. Within the private sector, trade and fishery formed the two major areas of economic activity. However, free enterprise was so enticing that both garrison officers and civil administrators actively engaged in both trade and the fisheries, and, with the private contractors, may be considered the principal recipients of any government largesse.

Given Isle Royale's short and turbulent history, it is not surprising that the colony's population remained relatively small during both French periods. As seen in Table 1,[4] the colony's resident population never exceeded 9000 and towards the end of the first period it barely exceeded 5000. The population was also surprisingly urban in character with approximately 35% of the population concentrated in Louisbourg. It is important to note that this urban concentration was exaggerated by the inclusion of the colony's military garrison, of which over three-quarters were stationed in Louisbourg. Indeed, the tripling of this garrison in the second French period further exaggerated the colony's urban character. Although the garrison had considerable interaction with the private sector, with its soldiers as a labour force and its officers as entrepreneurs, it still remained largely separated from the private sector. Another important segment of the population was the large labour force employed in the fishery. The transience of this labour force led to large seasonal fluctuations in the population and meant that the resident *habitant-pêcheurs* or fishing proprietors had a produc-

Table 1. Colonial population, various years

| Year | Population (including military garrison) | | | |
	(1) Louisbourg	(2) Isle Royale	(3) Isle Royale plus Isle St. Jean	(1) as % of (3)
1719	853	ca. 2012	2262	37.7
1726	1296	3528	3950	32.8
1734	1616	3955	4527	35.6
1737	1963	4618	5181	37.8
1752	4174	5845	8814	46.9

tion capacity considerably greater than their numbers indicated at some points in the year.

The best documented, but also the most exaggerated, aspect of the colony's economy was public expenditure. In addition to the well-known expenditure on the colony's fortifications and military garrison, the French government also maintained the civil administration, including the courts. The crown also contributed funds for the construction and maintenance of other public works, such as the hospital, the lighthouse, and a careening wharf. The extent of this expenditure has been considerably magnified by some historians. The total cost of the fortifications has been placed as high as 30 million *livres* instead of the approximate actual cost of 4 million *livres*.[5] This latter amount was less than the sum of 3 years' production in the colony's fishery at local production values. In fact, during most years the amount spent on fortifications was less than the outfitting costs of a 6-month voyage for a large warship of the period.[6]

The colony's government usually operated on a balanced budget based on a system of controls to prevent abuse.[7] In the fall local officials prepared a statement of anticipated expenses for the coming year, which was sent to the Minister of Marine for approval. This statement was scrutinized during the winter and payment, in the form of either cash or goods, was sent in the spring to Isle Royale for the approved items. These payments then appeared under one of three headings on the colony's balance sheet. First there were charges for the labour and material used on the fortifications. Second there was the

colonial budget to cover the annual expenditures of the colony, such as the salaries and supplies for the garrison and bureaucracy. Finally there were extraordinary expenses which covered special or unforeseen costs. Table 2[8] shows the colony's annual receipts and expenditures for the years 1721-57. It should be noted that the colony's receipts were supplemented on occasion by the sale of government supplies to private individuals and to French military units operating in Acadia on a separate budget.

When compared with Canada, Isle Royale received a far larger budget than its population warranted. With a considerably smaller population, Isle Royale's budget was over half as large as that of Canada, some six to eight times as much in terms of per capita funding.[9] As Isle Royale was more dependent on trade than Canada, the island colony had a higher per capita balance of payments. Government expenditure helped equalize this balance of payments, particularly during the second period, when increases in government expenditure greatly outstripped increases in the private sector. It is important to remember, however, that a considerable sum in the colony's budget did little to employ local services or encourage domestic production. In addition to the cash sent for salaries and local purchases, a large portion of the budget was spent on goods in France. Although the shipment of these goods enabled the colony to support a larger government sector than would otherwise be the case, they added little direct wealth to the local private sector.

Like government expenditure, commerce developed into one of the economic mainstays

Table 2. The budget of Isle Royale: receipts and expenses, 1721-57

| Year | Receipts | | | | Expenditure | Surplus(+) or deficit(-) |
	Colony	Extraordinary	Fortifications	Total		
1721	151 871	11 084	80 000	242 955	242 954	
1722	124 740	4 020	80 000	209 661	192 353	+17 308
1723	144 289	6 817	130 000	281 105	267 761	+13 344
1724	151 485	9 601	150 000	311 087	298 831	+12 256
1725	116 941	3 960	150 000	270 901	270 899	
1726	136 911	8 879	150 000	295 701	295 790	-89
1727	144 889	14 939	150 000	309 829	309 790	
1728	--139 056--		150 000	289 056	286 746	+2 310
1729	--155 112--		150 000	305 112	292 324	+12 798
1730	154 283	4 007	152 700	311 162	311 162	
1731	149 965	5 067	128 900	300 427	300 427	
1732	167 362	420	128 900	296 682	296 682	
1733	179 784	583	130 335	310 704	310 703	
1734	179 441	575	128 900	313 587	313 586	
1735	209 091	492	128 900	338 484	338 481	
1736	205 389	2 437	128 900	337 370	337 370	
1737	216 012	1 133	128 900	346 045	346 044	
1738	215 123	218	128 900	349 455	349 455	
1739	--176 005--		128 900	304 905	309 904	
1740	224 586	2 892	128 900	355 830	355 845	
1741	247 314	5 284	128 900	380 701	380 702	
1742	232 269	4 974	128 100	365 346	365 345	
1743	352 650	14 709	128 100	495 461	495 468	
1744	335 825	83 553	128 100	547 480	547 436	+44
1749	1 082 569	6 241	48 420	1 137 231	1 194 724	-57 492
1750	851 478	532 634	143 200	1 527 312	1 463 086	+64 266
1751	846 791	89 761	28 400	964 952	1 369 560	-404 608
1752	1 184 095	350 259	80 000	1 614 354	1 305 355	+308 998
1753	422 035	349 938	51 720	823 693	892 834	-69 141
1754	456 300	208 693	82 000	806 993	960 907	-150 914
1756					1 069 574	
1757					1 113 691	

of Isle Royale. Although the available primary documentation prohibits evaluating in absolute terms the contribution of commerce to the colony's economy, it is possible to show its relative importance to Isle Royale by comparison with other colonies. In Canada and Isle Royale, a single staple product (furs and fish, respectively) dominated each colony's exports. The per capita value of total exports for Isle Royale in 1737 was approximately eight times greater than that of Canada for the years 1735-39. Indeed, a comparison between Isle Royale and the neighbouring British colonies for approximately the same period reveals a similar pattern.[10] Only Newfoundland, which had a similar dependency on the fishery, came close to Isle Royale's level of per capita exports. This does not mean that Isle Royale was more prosperous but it does reflect the greater importance of trade to the Atlantic colony. Isle Royale was also more dependent on imports for the necessities as well as the luxuries of life. The resulting high level of exchanges provided Isle Royale, and Louisbourg in particular, with international markets and supply sources and fostered a strong shipping and merchandising sector.

The fishery formed a strong nucleus around

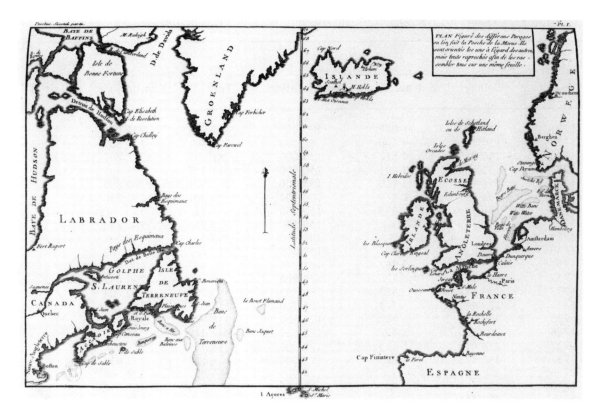

Figure 1. Eighteenth century plan of North American and European cod fishing grounds. (Source: Duhamel Du Monceau, *Traité général des pesches et histoire des poissons ...*, Part II, Section 1, **Sailant & Nyon** et Disant, Paris, 1772.)

which the rest of the colony's commerce revolved. This industry produced a large quantity of dried cod for export to markets in Europe and the West Indies. The fishery demanded a large material input, principally salt, and, as it was labour intensive, its workers placed heavy demands on the local economy for provisions and clothing. The French adopted a simple development strategy of minimizing these production costs through cheap imports. There were inherent risks in this economic system, however. Poor fishing seasons caused fluctuations in the general level of commerce and induced hardship. Similarly, interruptions of supply, such as crop failure or war, created shortages which adversely affected the whole economy.[11] It is important to note that these fluctuations and shortages were temporary in nature and until combined with an enemy attack, they never seriously threatened the colony's economic core. Indeed, the steady demand for Isle Royale's staple product, fish, gave the colony

a stability often lacking in other developing areas.

The fishery also aided Louisbourg's development as the colony's entrepôt. The dispersed pattern of fishing settlements along the island's Atlantic coast necessitated a distribution and collection centre. Such a centre maximized shipping efficiency by ensuring the sale and purchase of complete cargoes in one port equipped for a large turnover of goods. The outports, lacking the necessary volume of trade, were linked to the entrepôt through the *cabotage*, or coastal trade, which employed smaller vessels. Because of its early emergence as the largest population centre, its good harbour facilities, and its relatively central position in the colony, Louisbourg became the colony's commercial centre. Distance and market size slightly weakened its trading monopoly over the outports; Niganiche (Ingonish) and Petit Degrat, both relatively large and distant centres, were the only outports that regularly

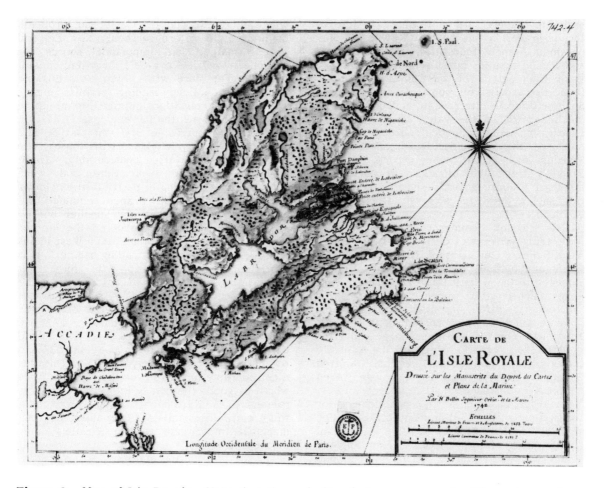

Figure 2. Map of Isle Royale. Note that the majority of place names and settlements are along the island's Atlantic Coast. (Source: Map Collection, Fortress of Louisbourg, No. 1742-4.)

received trading vessels from abroad. The scale of this commercial activity was small, however, and posed no threat to Louisbourg's supremacy.

Geography also enhanced Louisbourg's position as an entrepôt within the French imperial system. In an age when navigational instruments determined latitude but not longitude, it was common practice to sail along a selected latitude until a landfall was reached. Cape Breton's easternmost point was on the same latitude as the major French ports of Rochefort and La Rochelle, making Louisbourg a convenient destination for French ships sailing to North America.[12] Louisbourg was also used as a trans-shipment point in France's Quebec trade. By trans-shipping cargo at Louisbourg, vessels from France avoided the time-consuming and potentially hazardous

navigation of the St. Lawrence. Cargoes were efficiently carried from Louisbourg to Quebec in smaller vessels locally owned in the two ports. Louisbourg also benefitted as a pivotal spot in France's intercolonial trade. Although this trade incorporated triangular patterns for movement of goods between France, the West Indies, and Isle Royale, individual ships, like their British counterparts, probably engaged in a "shuttle" service between two of these points.[13] Louisbourg merchants extended this trade pattern to include trans-shipments to Acadia and New England as well as to Quebec.

France's Atlantic ports, especially St. Malo and St. Jean de Luz, were Isle Royale's most important trading partners.[14] These ports drew not only on a large and varied domestic production but supplemented these supplies

7

with re-exports from France's extensive foreign trade. In this way Isle Royale not only obtained French foodstuffs, clothing, wine, and fishing supplies but foreign goods ranging from Irish salt beef to Chinese porcelain. Dried fish and fish oil, some of which went to Spain and the Mediterranean, accounted for as much as 90% of the value of Isle Royale's export shipments. Lumber and coal made up much of the remainder. Distance had a limiting effect on some aspects of Isle Royale's trade with France; in particular, alternative sources for perishable commodities, such as fresh foods and livestock, had to be found.

Isle Royale's commerce also merged with the trading patterns of Canada and the French West Indies. Canada sent Isle Royale foodstuffs and lumber, some of which were re-exported to the West Indies. The French West Indies shipped return cargoes of sugar, sugar by-products, coffee, and other tropical goods, some of which could be passed on to Canada. Trade with Canada was never large and after the late 1730s, crop failures, growing domestic consumption, and other events limited Canada's exports to Isle Royale even further.[15] In contrast, Isle Royale's West Indian trade, based on a reciprocal demand for goods, flourished. Cod accounted for 70-80% of Isle Royale's exports to the Caribbean. Timber, coal, and re-exports of New England products, notably horses during the second French occupation, added variety to the trade. Rum and molasses formed the bulk of Isle Royale's West Indian imports. Such products filled a strong local demand such as quenching that common predilection of North American seafarers for rum, as well as forming the basis of Isle Royale's New England exports.

Acadia and New England were Isle Royale's two remaining trading partners. These areas, both under British control during this period, bridged Isle Royale's trade gaps within the French colonial system. Both regions provided perishable foodstuffs and livestock which, because of distance and other factors, were largely unavailable from French sources. Isle Royale's Acadian trade was principally with the French Acadian farmers whose self-sufficient farming economy restricted the nature and extent of commerce. This trade remained small in scale and consisted of foodstuffs, livestock, furs, and some fish being exchanged for manufactured items. New England's exports were similar, with the important addition of construction materials,

whereas re-exports of Caribbean products dominated the return cargoes. New England also remained an important source of schooners for Isle Royale's fishery and coastal trade.[16] The ease with which New England merchants obtained trading permits during both French occupations did much to minimize smuggling between the two regions.

As shown in Table 3,[17] the nature and extent of Isle Royale's trade with these three different regions varied considerably. These variations changed over time and were affected by a growing disparity in the colony's balance of payments. France remained Isle Royale's largest market and supplier but by 1754 this trade greatly favoured the mother country. After France, the French West Indies was Isle Royale's largest trading partner. It is particularly important to note the great increase in this branch of trade during the second French occupation. A similar growth pattern occurred in the colony's commerce with New England. Although data for 1754 are missing, Isle Royale's imports from and exports to New England stood at 488 037 and 654 680 *livres*, respectively, in 1752. This was a considerable increase over the 102 198 *livres* worth of goods the colony imported from this region in 1737. Trade with Canada and Acadia was relatively small and remained of only marginal importance to Isle Royale's commerce as a whole.

In the prosecution of these various trades, Isle Royale received visits from a substantial volume of ships. J.S. MacLennan concluded that after eliminating local traffic an annual average of 154 vessels visited the colony's ports, particularly Louisbourg, during the decade 1733-43. Only three ports in the more populous British colonies to the south surpassed this total.[18] Variations in average vessel size for the different trades make tonnage figures a more accurate reflection of the volume of trade with each area than a straight vessel count. During the first French occupation the annual volume of shipping amounted to some 8000 *tonneaux*.[19] The larger vessels from France usully accounted for less than half the number of ships but well over half the tonnage. During the early part of the second French occupation, the volume of shipping jumped to 13 000 *tonneaux*. The increase was due to a quadrupling of the West Indian and New England tonnage to parity with that of France.

On the basis of Isle Royale's export dependence on the fishery, Christopher Moore has

Table 3. Values of Isle Royale's imports and exports in *livres*, by region, 1737 and 1754

Region	Imports (*livres*)		Exports (*livres*)	
	1737	1754	1737	1754
France	1 022 597	1 437 256	1 082 394	788 757
West Indies	247 049	1 188 917	147 828	656 353
New England	102 198	n.d.	n.d.	n.d.
Canada	23 851	39 607	72 855	n.d.
Acadia	22 994	n.d.	n.d.	75 575
All regions	1 418 860	2 665 780	1 499 448	1 510 685

determined that the colony's balance of payments remained favourable until the late 1730s.[20] After that date a decline in fishery production resulted in a weakening of Isle Royale's trade position. During the second French occupation higher prices partially affected lower fishery production, but the increase in population and consequent increase in import demands further worsened the colony's balance of payments. Indicative of this decline in Isle Royale's trade position was an apparent tightening in the colony's money supply. Although bills of exchange were a major form of financial transaction, there were still considerable amounts of specie in circulation. Although it is impossible to quantify changes in this system, literary sources indicate the availability of specie to all classes during the first French occupation. By the 1740s the colonial administration was expressing concern over the increasing scarcity of cash.[21] Although Isle Royale was not reduced to issuing card money like its sister colony of Canada had to on occasion, the decrease in the money supply indicated a cash outflow to pay for goods and services.

The outflow of cash from the colony was aggravated by the French reluctance to diversify its economy and thus substitute domestic production for imports. Although French mercantilist policy prohibited local involvement in some areas of production, other areas, particularly in primary industry, were open for expansion. Certainly the colony had potential for greater development in agriculture, lumbering, and coal mining. A hundred years later, the Scots, utilizing a farming technology similar to that of the French, established an extensive subsistence agriculture.[22] The English were quick to set up a colliery at

L'Indienne (Lingan) during their brief occupation from 1745 to 1749.[23] The local construction industry provided a ready market for lumber while the colony's fishing, coastal, and even international trades provided a steady demand for vessels.

Numerous reasons have been advanced to explain in part the reluctance of the French to exploit more fully these economic opportunities. The French preference for grain over the more regionally suited potato hindered the development of farming beyond mere garden plots.[24] Competition with foodstuffs imported from France, New England, Quebec, and Acadia further restricted agricultural development. Similar competition existed with regard to construction materials and shipbuilding. During the first French occupation coal mining was relatively small scale, with local consumption seemingly limited to the artillery forge and presumably some private forges, plus small amounts for export. After 1749, however, coal started to become a major fuel and the English coal yard, on the eastern end of the quay, was retained by the French.

Indeed, too often historians have expected to see major development in what was essentially a new colony. Like all French colonies, Isle Royale suffered from the general lack of emigration from the mother country.[25] Although this was offset to some extent by large seasonal influxes of transient fishermen, there is nothing to indicate that these fishermen formed an effective pool of potential settlers. Given a slow immigration rate, the colony's short and turbulent history militated against any substantial development or diversification of the economy. After all, Isle Royale enjoyed barely three decades of peace before it was captured and its population deported in 1745.

There are indications, however, that had the French enjoyed a longer second occupation after 1749 there would have been a greater diversification of the colony's economy. The influx of Acadian settlers into Isle St. Jean and even into the interior of Isle Royale would have inevitably led to a greater agricultural self-sufficiency. Similarly, the government-encouraged settlements of Rouillé and Village des Allemands on the Mira River, while hardly successful, indicated increased government interest in other sectors of the economy. As in agriculture, the French also increased their exploitation of the colony's coal resources during the second occupation. However Isle Royale's truncated history precludes any analysis of its economy except on a short-term basis.

Within the short term the concentration of production factors within the fishery resulted in an economy dominated by the staple production of a single commodity — dried cod. Although this concentration left Isle Royale exposed to the fluctuations of international supply and demand, it nevertheless provided the colony with an export item generally in demand in the world market. With the fishery providing a cash income, Isle Royale was able to pay for its necessary imports and thus avoided the "starving time" that initially plagued so many developing colonies. This did not mean that there were not fluctuations in the fishery that caused occasional hardships. With so much dependence on the success of the fishery, fluctuations in the local level of production or in its market value or in the prices of imports had damaging effects on the colony's economy.

THE FRENCH FISHERY AND THE
ESTABLISHMENT OF ISLE ROYALE

The beginnings of the French cod fishery in the northwestern Atlantic considerably predated the founding of Isle Royale in 1713. Indeed, a French Basque memoir dated 1710 claimed that the Basques had been the first to exploit the North American cod stocks and that it was the acquisition of a Basque "rutter" or pilot's log book that had enabled Columbus to undertake his discoveries.[1] Initially drawn to North America in pursuit of whales, the Basques quickly took advantage of the abundant cod stocks — first for shipboard and domestic consumption and later for export. Although the archives attesting to these facts have been burned, the Basques maintain that North American coastal nomenclature supports their claims. Although the early date mentioned in these Basque claims may have been exaggerated, recent historical research has verified the existence of a viable Basque whale fishery in the Gulf of St. Lawrence and along the Labrador coast from the 1540s to the end of the century.[2]

Basque participation became one of the continuing trends in France's North American cod fishery. Both Nicholas Denys and Duhamel Du Monceau described the Basques as being the most skillful of the French fishermen. This Basque participation resulted in their presence at Plaisance and later at the founding of Isle Royale. Michel Daccarrette was one such Basque who had initially settled at Plaisance and had joined the wholesale transfer of that population to Isle Royale where he became a prominent fishing proprietor and merchant.[3] St. Jean de Luz annually outfitted numbers of trading and fishing vessels for Isle Royale, and Basque merchants, like Bernard Detcheverry, operated from this port and were a common feature of the Louisbourg business community.[4] Similarly, skilled Basque fishermen found a ready market for their services among the colony's fishing proprietors. The numerous entries of Basque names in the account books of the widow Dastarit, an innkeeper in Louisbourg's Fauxbourg district, confirms the frequency of this practice.[5]

Like the Basque monopoly on the whale fishery, which disappeared with the development of the Spitzbergen whale fishery in the 1620s, the initial Basque hegemony in the northwestern Atlantic cod fishery vanished with European exploration of the region. By the middle of the 16th century all the major western European powers — France, Spain, Portugal, and England — were engaged in this fishery. Thereafter the international competition for superiority would continue to the present day. As the intricacies of this struggle have been chronicled numerous times,[6] it suffices at this point to add that by the beginning of the 17th century the two main competitors were England and France.

The reasons for this struggle were both economic and political in nature. Economically, dried or salted cod became a major foodstuff in Europe and later in the colonies, so that its production was profitable. Moreover, the fishing industry became a source of consumption for domestic products while fish surpluses could be exported abroad to pay for the country's necessary imports. In this manner, the cod fishery became almost a textbook example of the emerging mercantilist model. Strategically the cod fishery trained large numbers of fishermen/sailors and so became the famed "nursery of seamen" for use in national navies during wartime. At a time when European expansion overseas emphasized the need for naval strength, the strategic implications of the cod fishery could not be overlooked.

All of these economic and strategic considerations were based on the cod's suitability as a preservable food item. The deterioration of fresh cod came from autolysis or self-digestion of the tissue by the enzymes and from putrefaction or bacterial decomposition.[7] Fortunately, both these conditions were easily overcome by reducing the moisture content of the flesh either through salting or drying, or a combination of the two.[8] In any case, the whole or "round" cod was first "dressed," i.e. the head and entrails were removed; the backbone was split about a third of the way up from the tail and removed so that the fish lay flat; and usually the black membrane lining the abdominal cavity was rubbed away to produce an attractive finished product. In Iceland and Norway the fish was simply exposed to cold weather drying which produced the famed hard-dried, stock fish. In more southerly regions, where warm weather speeded decomposition, salting was required in addition to air drying. Cod could be preserved solely by a heavy salt pickle or could be salted and then dried. The longer the fish was kept salted before drying the more salt was required. As salt is hygroscopic, heavily salted fish could only be dried to a certain extent before they absorbed moisture from the air. Consequently, lightly salted fish could be dried "harder" than the heavier salted ones. Generally speaking, lightly salted hard-dried cod was the preferred cure on the international market.

The difficulties that attended the curing of good quality dried fish were legion.[9] Failure to bleed the fish and rough or excessive handling encouraged bacterial decomposition. Improper splitting and failure to remove the abdominal lining resulted in a less attractive product. Insufficient salting failed to retard decomposition and excessive salting burned the fish. Mineral impurities in the salt imparted a bitter taste to the fish, delayed the penetration of the salt into the fish, and drew moisture from the air to the finished product.[10] The presence of red haliophilic bacteria in solar salt caused "reddening" of the finished cure, and the presence of brown mold caused eventual putrefaction.[11] Although it is uncertain if the above were present in 18th century supplies of French solar salt, they certainly plagued 19th century fishermen. If the fish dried too quickly a salt crust formed on the surface and if they dried too slowly the fish became "slimy."[12] Slow drying also caused putrefaction, which allowed soft spots to develop in the thicker parts of the fish.[13] On hot days the drying fish could become "sunburnt," with the protein of the fish coagulating like the boiled white of an egg. In general it must be remembered that curing salt fish was a highly skilled art and that any error in the dressing, salting, or drying processes lessened the value of the finished product.

Naturally, given the long history of the fishery and differences in fishing practices, a specialized nomenclature has developed to identify distinct types of fisheries. The most familiar related pairings in this terminology included green and dry, shore and bank, boat and vessel, and migrant and resident. These terms referred respectively to differences in the processing of the catch, the fishing grounds used, the fishing equipment employed, and the labour force engaged. In addition, the term sedentary referred to the prosecution of

the fishery from a permanent shore establishment by either resident or migrant fishermen. As the type of fishing gear or the kind of processing of the catch was frequently related to the fishing grounds being used, single terms of the specific pairs were occasionally used interchangeably. For example, the terms green, bank, vessel, and migrant might be used individually to describe a fishery that was actually a composite of all the terms. Without advocating the use of cumbersome multi-adjectives, careful attention has to be paid to the specific meaning of terms describing particular fisheries.

The green fishery referred to the preservation of fish solely through heavy salting whereas the dry fishery, at least in its North American context, referred to a technique by which the fish were first salted and then dried. With abundant and cheap supplies of domestically produced solar salt the French had a relative advantage over the English in the green fishery.[14] French fishermen also had the advantage of a large Catholic market for salt cod which had to be principally satisfied through the green fishery because of the earliness of the Lenten season.[15] As green salted fish gained consumer acceptance, a large and continuous domestic demand for the product, particularly in Paris, proved a mainstay for the French green fishery.

The English, with a smaller domestic salt supply and a greater reliance on exports, concentrated their efforts on the dry fishery, which required lesser quantities of salt. The dry fishery necessitated shore establishments, however, so the English forcibly advanced their claim to the exclusive use of the Avalon peninsula in Newfoundland.[16] Their success in advancing these claims has suggested that the English concentrated on the dry fishery and the French on the green. However, the French were also concerned with the dry fishery, as shown by their establishments at Plaisance and later at Isle Royale and by the tenacity with which they lobbied for shore drying rights in the repeated peace negotiations with Britian during the 18th century. The reason for the French interest in the dry fishery was the better preservative qualities in warm climates of dried over green cod. This made dried cod the more desirable commodity for export to the Iberian, the Mediterranean, and later the West Indian markets.

The particular fishing ground used was often directly related to the method of processing the catch. Within the northwestern

Atlantic cod fishery there were two types of fishing grounds — the inshore and the bank. The inshore fishing grounds hugged the coastline and were generally within a dozen miles (1 mile = 1.609 km) of the shore. The bank fishing grounds were actually the submerged plateaus of the continental shelf and were usually 50 miles or more from shore. They included Grand, Green, and St. Pierre banks off Newfoundland, the western banks off Nova Scotia, and Orphan and Bradelle banks in the Gulf of St. Lawrence, as well as the bank surrounding the Magdalen Islands. As no contact with land was needed, the green fishery supported fishing trips of several months duration to the offshore banks until the requisite amount of fish was caught. The dry fishery lent itself to the exploitation of the inshore fishing grounds with fishermen making daily trips to and from inshore grounds. A combination of these two fisheries also existed in which a relatively long trip of perhaps a month would be made to the banks, after which the heavily salted catch was brought to land for drying.

Understandably, the type of fish processing and the fishing grounds utilized determined the selection of the boat or vessel employed. The green fishery on the offshore banks favoured the use of larger, more seasonal vessels. The inshore dry fishery encouraged the use of small boats or shallops as the fish were taken ashore daily and carrying capacity was less of a consideration. In addition, a number of small shallops had a better chance than a single vessel in finding the dispersed schools of cod. The prosecution of a dry bank fishery necessitated the use of an intermediate-sized vessel. The fore-and-aft rigged schooner or *goélette*, averaging some 35-50 tons (1 ton = 0.907 tonne), proved large enough for the successful exploitation of the bank fishery yet small enough to make the relatively frequent trips to shore without underutilization of cargo space.

Theoretically, both resident colonists and migrant seasonal fishermen from the mother country could participate in all branches of the cod fishery. However, because the catch of the green bank fishery was transported directly to the home market on board the fishing ship, the migrant vessel fishermen probably had the advantage. Participants in the dry fishery often produced a small quantity of green fish towards the end of the season while waiting for final processing of the last batch of drying fish. Both resident and migrant fishermen conducted the dry fishery using either shallops or schooners. As the cod fishery did not usually begin until the end of April or the beginning of May, migrant fishermen easily arrived in the colony in time to make the necessary shore preparations for processing the catch. Permanent residents had an obvious advantage in selecting the best shore properties, an advantage that led to bitter rivalries between migrant and resident fishermen.

The sedentary fishery was closely associated with the development of a resident fishery but was actually dependent on possession of legal title to the land. This fishery usually included dry processing and was conducted from a permanent shore establishment. This procedure was ideally suited to colonial residents and contrasted sharply with the usual land division on a "first come first serve" basis as practised in the migrant fishery. However, there was the danger in the sedentary fishery that legal title to the land would pass into the hands of migrant fishermen who would occupy the land only seasonally. Without some control on non-resident land tenure, colonial development could be greatly hindered.

As might be anticipated from these concerns over the sedentary fishery, the various branches of the cod fishery had differing effects on the development of the North American seaboard. The French green vessel fishery conducted on the offshore banks had no land contact at all, aside from the infrequent trips inshore for fresh water or to repair storm damage. The migrant dry fishery resulted in only marginal colonial development as occupation was only seasonal and French-based fishermen tried to discourage competition from resident fishermen. The resident dry fishery provided permanent settlement but it did not encourage economic diversification. As the dried cod represented the finished product, no additional economic activity was undertaken except in the marketing sector. Although the fishery provided a market for goods and services, some items, like salt, and in some areas, food, were not produced locally. Even in those instances where colonial production was feasible, cheap imports from established areas and the limited size of the market often mitigated against the local firm. Often colonial production was limited to vessel and boat building and the strongest economic stimulus was to the service sector, supplying the labour force and maritime-oriented trades such as coopers and cordwainers.

Although a typically high level of imports, as in 18th century Isle Royale and 19th century Newfoundland, discouraged diversification of domestic production, the large volume of dried cod exports encouraged a strong merchandising sector. As the fishing outports were quite independent of each other in terms of actual dried fish production, the merchandising sector provided economic bonds within the colony. The scale of operations of individuals and companies affected the strength of these bonds. Large-scale operations such as those conducted by the Jersey firms in 18th and 19th century Gaspé and Cape Breton meant that trade links with the mother country were stronger than those with other outports in the colony. Small-scale operations prevented such business harmony and encouraged the development of a local business community. However, the typically small size of outport markets and production worked against the establishment of direct trade links in every outport. Instead, colonial entrepôts developed where the scale of operations warranted direct trade, with the movement of goods to and from the outports being conducted through the coastal trade. This system, which was practised in Isle Royale, ensured that an outport's direct economic bonds were to the colonial entrepôt rather than to its neighbouring outports.

Bearing these considerations in mind, it is necessary at this point to review briefly the development of the French fishery up to the founding of Isle Royale in 1713. After the beginning of the 16th century there was a gradual expansion in the northwestern Atlantic cod fishery and in the development of the international fish trade. France, with the advantage of a large domestic market, was able to concentrate on the bank, green fishery. Indeed, this extensive fishery remained a continued feature of France's cod fishery throughout Isle Royale's existence.[17]

At the same time, the growth of the international dried fish trade encouraged France's participation in the inshore dry fishery. By the middle of the 17th century the French dry fishery was divided between two regions in Newfoundland. The "Petit Nord" stretched along the west coast of the Great Northern Peninsula, and the coast of Chapeau Rouge eventually stretched from Plaisance to Cape Ray but the fishery was concentrated along the Burin Peninsula.[18] In Acadia, development of an extensive migrant fishery was somewhat hindered by a less abundant inshore fishery, a further distance from France, and the occasional vexations caused by the claims of the land proprietors.[19] The French also conducted the dry fishery on the Gaspé peninsula and on the Labrador coast.

More important to the establishment of Isle Royale was the founding of Plaisance (Placentia) in 1662. At Plaisance the establishment of a resident fishery subsequently protected by a small garrison formed a blueprint for the later development of Isle Royale. A small French resident fishery on the south coast of Newfoundland between Trepassey and Cape Despoir had actually preceded the official founding. In 1660, Nicholas Gargot was chosen as the first governor of Plaisance. The development of the nascent colony was slow as residents had to fight stiff competition from migrant fishermen and the efforts of the colonial officials and garrison officers to monopolize the beach areas of Plaisance. Probably as the result of superior fishery facilities, the French resident fishery concentrated at Plaisance. In 1687, 256 of the 640 French inhabitants on the south coast were at Plaisance, which was protected by a garrison of only nine soldiers.[20] War with England during 1688-97 and again during 1702-13 prevented the development of this resident fishery and led to the colony's demise. By the treaty of Utrecht in 1713, the French ceded Plaisance to the English but were confirmed in their possession of the Petit Nord in Newfoundland and of Cape Breton Island.

The French swiftly moved to transfer their settlement from Plaisance to the new colony of Isle Royale, and by early September of 1713 the initial group of 149 men, women, and children had made the transfer. Efforts to broaden this population base by tempting large numbers of French Acadians to forsake the new British regime in Nova Scotia met with only marginal success. The agrarian background of the Acadians did not lend itself to easy absorption into a maritime economy, and the agricultural prospects of Isle Royale, even when coupled with promised freehold land tenure, failed to induce many to emigrate. This early unsuccessful attempt at diversification proved an accurate forerunner of the fishery's supremacy in the new colony.

Having had experience with sedentary fisheries in Gaspé, Acadia, and the Petit Nord as well as at Plaisance, the French Ministry of Marine quickly set up a legal framework for a resident fishery at Isle Royale. Initially the former residents of Plaisance were to be offered properties in Louisbourg, while the migrant fishermen had access to the beaches at Mira and Scatary.[21] The concessions made to the new inhabitants were to be based on their former properties in Plaisance and in proportion to the number of shallops they owned. Married fishing proprietors, whose families would ensure the colony's development, became the object of French official protection. As abuses on the ownership of fishing properties became apparent, the regulations respecting land tenure became increasingly restrictive. For example, single fishing proprietors were forbidden to rent their properties and migrant fishermen were forbidden to conduct the sedentary fishery at all.[22]

Similarly, the areas of commercial activity open to migrant fishermen and merchant captains were restricted to protect the fishing proprietors. An ordinance passed in 1720 required captains of all vessels arriving in Louisbourg to make an exact declaration of the cargo on pain of confiscation.[23] They could sell fishing gear and provisions to anyone, provided they informed the authorities, but were allowed to sell liquor only to merchants or fishing proprietors and not to tavern keepers. Merchants engaged in the fishery were able to retail liquor directly to their own employees only. A later regulation prohibited foreign merchant captains from engaging the fishermen of fishing proprietors and from purchasing supplies from other vessels.[24] This latter measure was designed to limit competition between residents and non-residents and thereby keep prices down. Later, general regulations allowed foreign merchants to sell only aboard vessels, prohibited captains from leaving fishermen in the colony to undertake the autumn fishery, and also attempted to prevent captains from buying the cargoes of vessels from France, Canada, or the West Indies.

Regulations were also passed that directly protected the financial position of the fishing proprietors. In 1743 an ordinance was passed that provided for the comprehensive regulation of the Isle Royale fishery.[25] The wages paid to hired fishermen were specified to eliminate expensive competition for labour, not to provide a minimum wage for the fishermen. Similarly, the hired fishermen were made liable for unnecessary damages to fishing equipment and for any good fishing days lost through negligence on their part.

In addition to this legally fostered resident fishery, migrant vessel fishermen operated from rented shore properties or from temporary establishments on land that was not formally conceded. It was the transfer of France's migrant as well as resident fishery on Newfoundland's south coast to Isle Royale that gave the nascent colony such a great impetus in dried fish production. As shown in Figure 3, the colony's dried fish production had already reached its highest level by 1718-19. Unfortunately, the shortness of the time period and gaps in the statistical material prohibit the use of more sophisticated quantitative techniques in determining trends. However, even a simple bar graph as in Figure 3 is useful in determining levels of production in some periods. Indeed, three periods emerge: one of relatively high production from 1718 to the late 1730s, marked by annual fluctuations and perhaps depression in production in the 1720s; a second of markedly declining production in the 1740s; and a final period of low production during the second French occupation.

The first of the two obvious periods of lower production took place during the 1740s. A decrease in the per unit landings of both shallops and schooners amplified a reduction in the actual numbers of these vessels employed. This decrease in per unit production was particularly severe in the winter shallop fishery which dropped from a high of 180 *quintaux* (a measure of weight equal to 100 *livres* or 48.95 kg or 100 lb.; the English quintal or hundredweight weighs 112 lb. or 50.97 kg) per shallop in 1739 to a low of 30 *quintaux* per shallop in 1743. The reduction in the number of shallops and schooners operating in the Isle Royale fishery became more acute prior to the outbreak of war in 1744, as pre-war tensions caused French outfitters to cancel voyages. Any such cancellations had a twofold effect on the colonial fishery as fishing proprietors encountered shortages of transient labour and supplies from France. A similar pattern of pre-war reductions in dried fish production has also been noted in Newfoundland during the 18th century.[26]

The second period of low production in the Isle Royale fishery occurred during the 1750s. Per unit production in both the shallop and schooner fisheries reached the high levels of the 1730s, but total volumes remained low. Initially, the resident fishery remained low as the fishing proprietors had to overcome the difficulties of re-establishing their operations. Within a few years there was a marked expan-

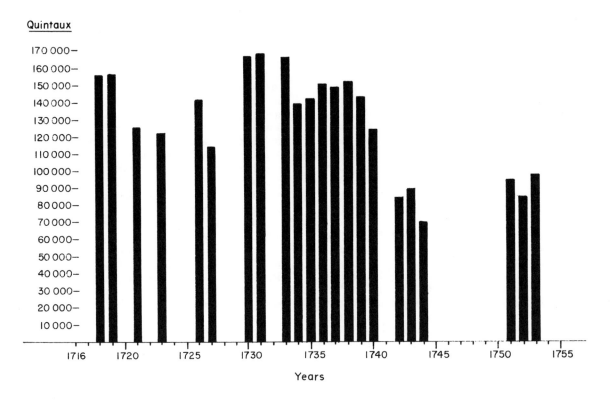

Figure 3. Export volumes of the Isle Royale cod fishery, 1718-55 (in *quinteaux*). (Source: Table 10.)

sion in the resident fishery, which was quickly reversed as the threat of war again loomed in the mid-1750s. The failure of the migrant fishery in Isle Royale to reach its former level of participation was the decisive factor in determining the low levels of production in the 1750s. This lower level of migrant participation also had a restrictive effect on the resident fishery. In 1750 the *commissaire-ordonnateur*, Prevost, reported that a labour shortage was limiting the fishing proprietors' catch. The decline in the number of migrant fishing vessels also reduced the amount of transportation available to transient fishermen who worked seasonally in the colony. The outbreak of hostilities in 1755 ended any chance for further development of the French fishery in Isle Royale.

With the exception of these two periods of low production, the Isle Royale fishery had an annual output of between 120 000 and 170 000 *quintaux* of dried cod. Although production was lower during the 1720s than in the years before and after, the statistical gaps for this decade make conclusions of a depression in output somewhat tentative. A more important trend, which is investigated more fully later in this paper, is the expansion of resident over migrant dried fish production. The total domination of the Isle Royale fishery by residents during the second French occupation contrasts with the strong migrant input prior to the first siege. Leaving this trend aside, the major factors influencing the annual fluctuations in the colonial industry were shortages of labour and supplies, price variations, and weather conditions.

The dominant position of the fishery within

Table 4. The "in France" value of the Isle Royale fishery in *livres*

Year	(1) Cod (*livres*)	(2) Cod oil (*livres*)	(3) Magdalen oil (*livres*)	(4) Total fishery (*livres*)	(1) as a % of (4)
1716					
1717					
1718	3 130 000	313 000		3 443 000	90.9
1719					
1720					
1721	2 512 000	168 000		2 680 000	93.7
1722					
1723	3 029 000	183 920		3 212 920	94.3
1724					
1725					
1726	2 818 000	140 900		2 958 900	95.2
1727	2 293 000	91 680		2 385 280	96.1
1728					
1729					
1730	3 312 600	165 600	12 000	3 490 200	94.9
1731	3 350 800	167 550	35 000	3 553 350	94.3
1732					
1733	3 307 300	181 830	11 000	3 500 185	94.5
1734	2 796 200	153 780	12 100	2 962 080	94.4
1735	2 849 900	156 750	13 200	3 019 850	94.4
1736	3 022 200	166 210	14 300	3 202 710	94.4
1737	2 986 000	164 230	22 000	3 172 230	94.1
1738	3 049 400	189 640	22 000	3 239 040	94.1
1739	2 873 200	157 960	30 250	3 061 465	93.8
1740	2 463 000	147 780	19 200	2 629 980	93.6
1741				2 585 440	
1742				1 782 680	
1743	1 774 400	106 440	42 000	1 922 840	92.3
1744	1 388 600	83 280	9 600	1 481 480	93.7
1750	1 811 200	108 660		1 919 860	
1751	1 911 600	114 600		2 026 200	94.3
1752				1 771 960	
1753	1 969 000	98 450	17 000	2 083 500	94.5
1754				2 054 075	
1755					

Isle Royale's economy cannot be questioned. In an export-oriented economy the fishery accounted for almost 90% of the colony's total exports in 1737 and for almost two-thirds of the total in 1754.[27] This was particularly important in light of the fact that much of Isle Royale's remaining exports actually consisted of re-exports of goods from other areas. If exports only of domestically produced goods were included, the dominance of the fishery would have been nearly absolute.

Within the fishery, production was almost entirely concentrated on one item — dried cod. Other food fish, such as herring or mackerel, were used only as bait and were not prepared for export, although they were the object of extensive domestic French fisheries. As seen in Table 4, dried cod accounted for over 92% of the estimated value of Isle Royale's fishery during both French occupations.[28] When the value of cod oil, a by-product of the cod fishery, was added to that of dried cod, the concentration of this single fish commodity became even more apparent. Indeed, the only competition to the cod fishery came from "Magdalen oil" derived from the killing of *vache marins* or walruses on the Magdalen Islands. The production of this oil never amounted to 1% of the total value of the fisheries in any year.

Care must be taken in using these figures, however. They were generally compiled by the *commissaire-ordonnateur* and were based on an estimate of the number of shallops and schooners fishing in the colony, multiplied by an average catch for each type, respectively. The amount of cod oil was then proportionately derived from the estimated total catch and a figure was also added for the Magdalen oil. The amounts of dried cod and oil were then multiplied by an official value for each commodity. It was at this point that the accuracy and utility of these statistics weakens, because the use of official values failed to reflect price changes. With one exception, dried cod was consistently valued at *20 livres* per *quintaux* and oil typically varied from 110 to 120 *livres* per *barrique*. Although their prices purported to reflect the value of these commodities in France, their value in Isle Royale was, of course, considerably less.

The Isle Royale value followed the current price along the coast according to the practice established at Plaisance. Although Peter Warren noted in a 1739 review of Isle Royale's fishery that the government annually set the price at about 9 shillings or 8 *livres* per *quintal*, this type of price control appears to have happened only once.[29] In 1724 St. Ovide and de Mésy set the price at 12 *livres* for fishermen selling to their fishing proprietors, but this probably represented a temporary measure.[30] Certainly, sales of dried cod at lower prices were recorded, particularly when the fish in question were not completely dried. It would appear that 10 *livres* per *quintal* represented an average price for the first French occupation. Consequently, by halving the values appearing in Table 4 an approximate colonial value for the fishery can be obtained. However, the use of constant official values has prohibited any meaningful comparison between price and the volume of fishery production.

The values of Isle Royale and Canadian staple exports can be compared, however, utilizing the adjusted evaluation of Isle Royale's fishery. In Table 5[31] the values of total

Table 5. Value of total exports from Canada and of fishery products from Isle Royale for various years (in *livres*)

Year	Total exports - Canada (*livres*)	Fishery products - Isle Royale (*livres*)
1726		1 479 450
1727		1 192 640
1729	1 483 198	
1730	1 398 327	1 745 100
1731		1 776 675
1732	1 483 192	
1733	1 389 047	1 750 092
1736	1 677 696	1 601 355
1739	2 103 868	1 530 732
1740	2 111 107	1 314 990
1753		1 041 750
1754		1 027 037

exports from Canada and of fishery exports from Isle Royale are compared. (In the case of Isle Royale's export values these have been reduced by half to approximate the colonial value of the catch.) Isle Royale's fishery exports compared favourably with the total exports of the more populous colony of Canada. It is important to note, however, that A.J.E. Lunn, who originally compiled the Canada figures, felt the amounts were underestimated. While higher per capita import costs offset commercial gains in Isle Royale, this small colony, on the basis of its fishery, had a combined import and export trade equal in value to the trade of the larger colony of Canada.

Just as the scale of its fishery assured Isle Royale of an important position within the French imperial trade system, it also gave the colony prominence within France's North American fishery. It has been estimated that the annual average of France's total northwestern Atlantic fishery effort may have reached 8 million *livres* between 1720 and 1760 and that Isle Royale accounted for 1.5 to 3.5 million *livres* of this total.[32] Although Isle Royale did not dominate French fishery production it did constitute a major source of supply.

The volume of Isle Royale's fishery production also made Isle Royale a significant competitor within the international cod fishery. In their estimates of the volume rather than the value of Isle Royale's fish production, the colonial administrator's reports were probably more accurate. Reports prepared in 1734 and 1735 by the Admiralty in Louisbourg provided greater detail with regard to the winter fishery but were otherwise in close agreement with those prepared by the *commissaire-ordonnateur*.[33] Similarly, there was a reasonably close concurrence between the stated exports of fish in 1737 and the amount of fish produce reported by the *commissaire-ordonnateur*.[34]

Using these figures on Isle Royale's fishery production, comparisons can be made with the fish production of the neighbouring British colonies. Between 1736 and 1739, when Isle Royale's fishery reached its height in volume of production, the colony produced an annual average of 149 120 *quintaux* of dried cod. For the same period the British fishery at Newfoundland, both shore and bank, produced an estimated annual average of 380 400 quintals. In 1745, when war had led to a decrease in production, New England still produced 230 000 quintals.[35] Although Isle Royale did not match the larger British colonies in terms of production it was nevertheless regarded as a serious competitor and a keystone of the French fishery in North America.

The fishery was Isle Royale's strength and its weakness. It stimulated the colony's export-oriented economy and enabled Isle Royale to become an entrepôt within the French imperial system. Even at colonial prices its value in the late 1730s was the equivalent of a subsistance wage for every man, woman, and child in the colony. In spite of fluctuations in production the fishery provided a fair degree of economic stability to the young colony. However, this concentration on the fishery detracted from the development of other sectors that would have broadened the economic base and increased self-sufficiency. The fishery concentrated the population in areas less suited for diversification into agriculture and lumbering. The labour pool attracted by the fishery was schooled in a maritime economy rather than a land economy. Perhaps most importantly, the fishery encouraged the importation of cheap supplies that kept the costs low but also provided stiff competition for fledgling domestic production.

LAND AND THE ISLE ROYALE FISHERY

With the development of the dry fishery land became an economic factor within the fishing industry. Vessels engaged in the green fishery had no real need for a shore establishment and hence land was not a cost factor. The dry fishery, on the other hand, necessitated a shore establishment for the cleaning, salting, and drying of the fish. This led to the construction and maintenance, even if only seasonally, of *chafauds* (stages), *graves* (beaches), and *vigneaux* (flakes). With so much of the fish processing done on land and with fishing largely restricted to a limited distance from these shore stations, living quarters such as *cabannes* and storage buildings such as *magasins* were also erected near the work areas. The investment of capital and labour into the improvement of these fishing establishments naturally increased their value and ultimately their maintenance costs.

Natural factors, even more than economic necessities, determined the desirability and suitability of a particular stretch of shoreline as a fishing station.[1] Perhaps the most important of these natural factors was the proximity of the shoreline properties to good fishing grounds and the shelter and accessibility offered to fishing vessels. Thereafter a whole range of other considerations apply: the area's closeness to adequate supplies of wood and fresh water; whether the shore was rock or beach; and the region's prevailing climatic conditions. For example, frequent fogs or excessive summer heat had a debilitating effect on drying cod. In short, not every stretch of coast was suitable for a shore establishment, and even within the selected areas the advantages of particular stations varied.

These variations in the value of shoreline land necessitated some method of alloting shore establishments to guarantee the orderly prosecution of the fishery. Within the migrant vessel fishery this problem was resolved by the establishment of fishing admirals. This post fell to the first fishing captain entering each harbour at the beginning of the season, a practice enshrined in custom and law. The fishing admiral had first choice of shore stations and to any salvage from the previous season. He also delegated the remaining lots to later arriving vessels and acted as the arbitrator of any dispute arising during that

season. Naturally the advantages accruing to the fishing admiral resulted in an undignified race every spring as each captain strove to arrive first in the harbour of his choice.

The eventual breakdown of this system came with the emergence of permanent resident fishermen in the seaboard colonies. These resident fishermen, as a result of their year-long association with a given locality, took the opportunity to secure the best shore establishments. The migrant fishermen were loath to concede them this right. Indeed, British fishing interests, using their financial power and political clout, managed to maintain some restrictions on settlement in Newfoundland until after the Napoleonic Wars.[2] Although the resident fishery never achieved a dominant position within the total French northwestern Atlantic fishing industry, provisions were adopted early to protect it. Although Nicolas Denys was permitted in his patent of 1654 to establish a sedentary or resident fishery in Acadia,[3] this fishery did not receive a substantial impetus until the French established Plaisance in 1662. In the following years a small but vital resident fishery was to develop there, which would later act as a pattern for the founding of Isle Royale.

Shortly after the colony's establishment in 1713, a legislative system evolved for the allotment of the valuable shoreline property, a system that gave preference to the resident fishery.[4] Initially the fishing proprietors settled on the unoccupied land of their preference but by 1718 most had ownership confirmed in official concessions. Thereafter aspiring fishing proprietors received grants of unconceded land through application to the governor and *commissaire-ordonnateur*. Having received legal title to his concession the resident could then rent or sell his property and hence repeated transactions required the proper marking of boundaries. Official concern over the use of shore properties by non-residents resulted in legislation favouring married settlers. For example, unmarried fishing proprietors were forbidden in 1723 from renting their concessions. Other legislation prohibited migrant fishermen from conducting a sedentary fishery, i.e. returning annually from France to prosecute a seasonal fishery from a permanently owned shore

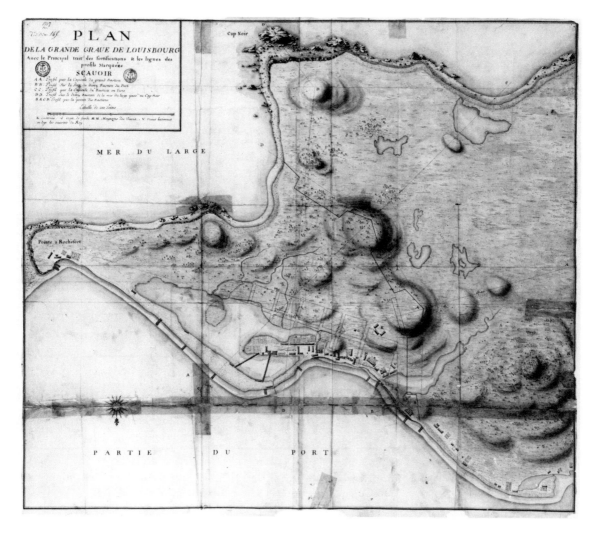

Figure 4. Plan of the Grande Grave (main beach) of Louisbourg. A block grid and fortifications trace were soon superimposed on the fishing properties scattered along the shoreline. (Source: Map Collection, Fortress of Louisbourg, No. 1717-2.)

establishment. Migrant fishermen either had to make temporary establishments on unconceded or unoccupied land or rent existing ones from married residents.

The value of land in the dry fishery and the fishing industry's economic primacy were reflected in the colony's settlement pattern. On the island of Isle Royale, French settlements were almost entirely limited to the eastern coast, stretching from Niganiche (Ingonish) in the north to Petit Degrat in the south, in harbours convenient for the fishery. The west coast, facing the Gulf of St. Lawrence, remained virtually uninhabited and the interior was only lightly settled during the 1750s. Settlement, and hence the fishery, was relatively restricted on Isle St. Jean during the first French occupation for a variety of reasons. Only one fishing and farming community, St. Pierre du Nord, attained any size.[5] The influx of settlers during the second period was largely Acadian refugees, whose agrarian orientation did little to develop the island's fishery. In any event, the colony's short history prevented complete development of land use potential within the sedentary fishery.

The value of the shore establishments also

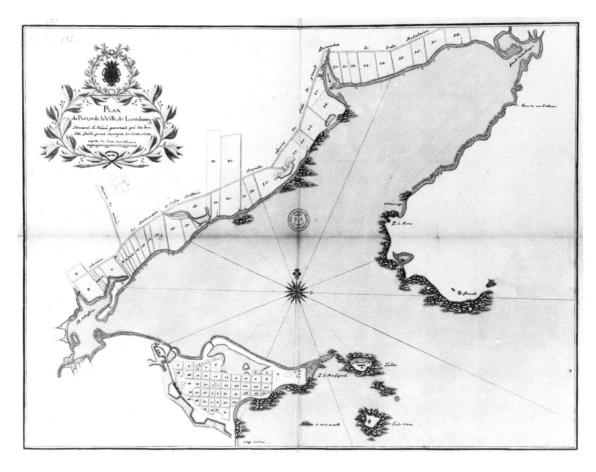

Figure 5. Plan of Louisbourg and harbour. Note the fishing properties laid out along the harbour's north shore. (Source: Map Collection, Fortress of Louisbourg, No. 1734-2.)

determined settlement patterns within the individual fishing communities. The *terrains de pêche* (fishing properties) ringing each harbour produced an elongated settlement similar to that produced by the narrow Canadian farms stretching back from the St. Lawrence River. Only in Louisbourg harbour was there any appreciable change over time in this development pattern. Early plans of the harbour show the fishing establishments stretching from Pointe de Rochefort along the future town's waterfront and around to Côte du Nord (see for example Figs. 4 and 5). As a result of the growth of both the fortifications and town properties, and perhaps because of the exposed position of Pointe de Rochefort, the shore establishments along the town's waterfront disappeared (see Fig. 6). By the mid-1730s the fishing stations were located only in the Fauxbourg between the Porte

Dauphine and the Barrachois de Lasson and along the Côte du Nord to the Fond de la Baye. Even there the location of the Royal Battery excluded a large section of the shoreline from utilization by fishermen.

As the concessions were based on the amount of land the fishermen actually utilized, fishing properties generally reflected the small scale of the owner's initial operations. In established fishing communities this made the later assemblage of large shore properties difficult or, at the least, expensive. Occasionally the small size of the fishing properties made it difficult to maintain minimum operations. François Malle de Laurembec in 1752 had built stages, flakes, and beach area for the use of two shallops in the coming season on a shore property he had purchased, but was uncertain whether the land would be sufficient.[6] In some instances fishing proprie-

22

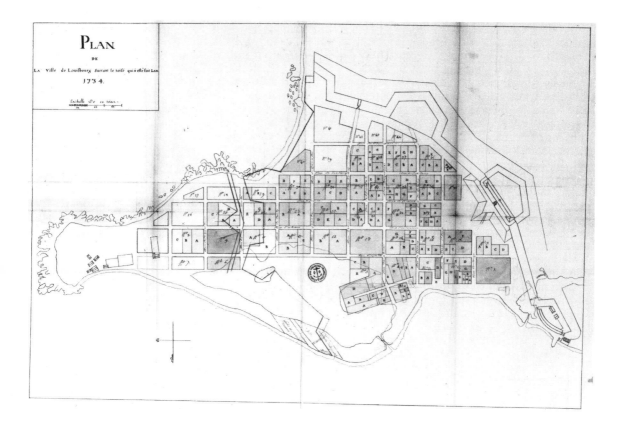

Figure 6. Plan of Louisbourg. By 1734 fortifications and a block system had replaced the earlier, irregular shoreline settlement. (Source: Map Collection, Fortress of Louisbourg, No. 1734-5.)

tors were able to piece together sufficient property. For example, Pierre Martissance received confirmation in 1734 of his concession on the Côte du Nord formed from pieces of three earlier grants.[7] Other fishermen were not so fortunate and had to contend with dispersed holdings. François Lessenne, for example, owned fishing establishments in Louisbourg's Fauxbourg and along the Côte du Nord, but the properties were not adjacent.[8] A greater scale of operations increased the need for dispersed holdings, as in the case of Michel and Jean Daccarrette, who were among the colony's leading fishing proprietors in 1726, with holdings in Niganiche, Saint Esprit, and Fourché (Fourchu).[9]

The importance of land in the dry fishery placed restrictions on what was otherwise an open access resource. As in agriculture, extensive land ownership could result in monopoly conditions. Efforts were made by prominent individuals during Isle Royale's existence to obtain just such extensive tracts.

The most notable example was the grant of the islands of Saint Jean, Magdaline, and Miscou made in 1719 to the Comte de St. Pierre.[10] Using the sedentary fishery as the basis of a large colonizing effort, St. Pierre claimed exclusive rights not only to the islands themselves but also to the extensive fishing grounds they bounded. The enforcement of this broad interpretation led to a legal dispute in the Admiralty Court of Isle Royale in 1721, which St. Pierre lost. Although he gained a reversal of the decision by the king's council the next year, continued difficulties and the expense of such a large colonizing venture forced him to virtually abandon his grant in 1725. Five years later St. Pierre's concession was annulled.

Other attempts were made to secure large grants of land suitable for the fishery but seldom on the grand scale of the Comte de St. Pierre. In 1719 Louis-Simon Le Poupet de la Boularderie received the post of commandant and a priority right to beaches needed to dry the produce of 100 fishermen at Niganiche, as well as land at La Petite Brador (St. Andrew's Channel).[11] The following year his shore rights were transferred to Isle D'Orleans (Ingonish Island) because of the overcrowding of the beaches on the mainland. Boularderie was unable to form a company to exploit the grant until 1723 and then, 3 years later, he dissolved the partnership in a dispute over policy. A new company was formed the same year but continuing managerial problems led to its dissolution in the mid-1730s. Thereafter Boularderie abandoned the fishery and concentrated his efforts on his grant at Petite Brador. Although his development efforts necessitated large investments and debts, Boularderie and his associates did achieve limited success with their large fishing establishments at Niganiche.

Not all attempts at assembling extensive shore property respected the technicalities of the law. In 1752 La Roque noted in his census of Isle Royale that Jean Hiriart and his partner, Pierre D'Aroupet, were operating an extensive fishery at Petit Degrat which occupied approximately half of the available fishing properties.[12] Unfortunately Hiriart did not have legal title to most of these concessions but had merely appropriated them. Indeed, so powerful was his personality or at least his means of persuasion that Larcher, a Louisbourg merchant operating an almost equally large fishery there, rented his fishing properties from Hiriart even though the latter did not own them. However, attempts to secure excessively large holdings, legally or otherwise, were relatively rare. Most highly successful fishing proprietors seemed to have preferred to enter the ranks of the *marchands* and *negoçiants* rather than expand their operations in a relatively risky fishery.

The importance of land as a factor of production was naturally reflected in the values placed on individual fishing properties. Such a property in an established community represented a substantial investment and rents and sales were set accordingly. For example, the heirs of Henri Nadeau *dit* Lachapelle rented his Fauxbourg property in 1743 for 420 *livres* per annum and 1 capon per *arpent* (approximately 1 acre).[13] This same property was evaluated in 1757 on the death of its then current owner at 7000 *livres*.[14] Great variations existed in the evaluations placed on different shore establishments. When Elie Tesson La Floury died in 1741 his Côte du Nord fishing station was valued at 6000 *livres*,

while his similarly developed site in Petite Brador was considered to be worth only one-tenth of that sum.[15]

These variations in property value reflected in part the improvements made on the particular site. While the Fizel property in the Fauxbourg was valued at 7000 *livres* in 1757, an adjoining but undeveloped fishing property had been purchased in 1752 for only 2600 *livres*.[16] Henry Nadeau *dit* Lachapelle spent a winter clearing and preparing a new fishing property at Fourché but had to return to his Fauxbourg establishment when the new site was still unfinished in the spring.[17] In 1751 Beaubassin Sylvain et Cie (a fishing company) hired two soldiers at the rate of 1 *livre* per *toise* to clear 265 *toises* of shore area for drying cod.[18] As 1 or 2 *livres* a day represented the usual rate obtained by soldiers in most civilian employment, the shore preparation must have been a lengthy process.

The need for this work resulted in higher rents for shore properties in Isle Royale than had been the case in Plaisance. In 1733 the merchants of St. Jean de Luz and Siborne complained that the habitants were charging excessive rents for their beaches.[19] They felt that as they rented only the beach area in Isle Royale the charge should be fixed at 10 *quintaux* of cod per schooner and 5 *quintaux* for shallops, as they had formerly rented beach, stage, and cabins at Plaisance for 1010 *quintaux* per shallop. Colonial officials dismissed this demand as unreasonable, noting that the beaches at Plaisance were formed naturally whereas those of Isle Royale required annual attention.[20]

The clearing of the property and the preparation of the beach represented only two aspects of the capital investment in buildings and work areas needed for the sedentary fishery. In 1726 Bonnaventure Le Brun described two well-equipped fishing properties he had developed, one at Petite Brador, the other at Niganiche.[21] The Petite Brador property included Le Brun's own dwelling which was 30 *pieds* by 20 *pieds* with two rooms, a kitchen, and two fireplaces. There was a storehouse, 24 *pieds* by 18 *pieds*, for storing drink and fishing equipment, with a storage bin of pine boards for holding bread. Another storehouse of the same size was used for salting cod and storing it when dry. A *pont* (bridge), 70 *pieds* by 5 *pieds*, led to a fish stage, 24 *pieds* square. There were eight flakes covering a total area of 50 *pieds* by 120 *pieds* on cleared land. A garden, 120 *pieds* square, was surrounded by a palissaded fence, 7 *pieds* high, with two gates. The cabin for the crews was *45 pieds* by 18 *pieds,* and half was covered by pine boards. There was also an oven, a chicken house, a shed for livestock, and a large yard.

The other property at Niganiche shared most of these features. There were two cabins which could house 10 crews, and a third housing the smith, carpenter, cooper, and five boys. One storehouse could hold 400 barrels of salt, while another, *45 pieds* by 20 *pieds,* had three partitions, a cellar, and a storage bin holding 200 *quintaux* of bread. A new cabin, presumably for Le Brun or his delegate, had two rooms and a kitchen. There were two gardens, one 40 *pieds* square, the other 80 *pieds* square and well fenced. The establishment also contained a chicken house, a sheep shed, and, perhaps most importantly, a bridge, 30 *pieds* by 5 *pieds,* to carry the cod ashore.

The descriptions of Tesson La Floury's fishing properties at Côte du Nord and Petite Brador in 1741 noted similar features.[22] At Côte du Nord, Tesson La Floury's establishment included his house, a storehouse with a storage bin, a stable, a beach, flakes, and cabins. His property at Petite Brador included two stages, a *mauvais magasin* joining a *magasin neuf* containing a storage bin, a *grand cabanne* with a storehouse at the end, three cabins for fishermen, a garden, and several flakes. Although these descriptions failed to provide much detail they nevertheless identified the common features of the shore establishments.

Among the more notable features were the fish stages that lined the shore and jutted out into the water. Their use as work stations in the preparation of cod for drying probably superceded any function they served in the transfer of other goods. Indeed, 18th century French writers used different terms for stages and wharves with specific functions. *Chafaud* appears to have referred to a stage fulfilling only this fishery function. Contemporary writers such as Denys and Du Monceau gave detailed accounts of both its construction and usage in the fishery,[23] and notations on early plans of Louisbourg often described the stages' use for the preparation of cod.[24] *Pont* seems to have been the term for stages used simply to transfer the catch to shore. Bonnaventure Le Brun had a *pont* at Niganiche for carrying the cod ashore and another at Petite Brador that led to a *chafaud*. Similarly, a sturdy wooden wharf built by Nicholas Larcher and designed more for the transfer of goods than

for use as a fish stage was described as a *calle* (wharf) rather than as a *chafaud*.[25]

Traditionally these stages exemplified a simple and, at times, flimsy construction. Denys and Du Monceau described in detail their construction from softwood poles.[26] At low tide the poles were set vertically into the deepest water attainable and were secured by angled props. Horizontal bracing further secured the numerous uprights and served as ladders on the sides of the stage. A floor of poles was then laid across the top so that it would be several feet above water at high tide. The whole structure was nailed together with large spikes, and *cloux de chafaud* were common. Naturally this type of construction required large numbers of trees and the road between Louisbourg and the Lorembecs was viewed in part as opening a new source of wood for flakes and stages.[27] Raymond's comments on the solid appearance of Larcher's wharf indicated that the stage standing on its numerous softwood piles did not appear overly sturdy. Although Denys and Du Monceau described the deck of the stage as ending in a triangular point, plans and views of such structures in Isle Royale show a simple square end.[28]

The deck of the stage was frequently covered with a simple roofed shed to protect the work area. These shelters covered most of the stage leaving an open area at the water end to receive the day's catch; the stages often had open walkways down one or both sides. In the migrant fishery these shelters were temporary, with sailcloth used to cover the roof and some of the walls. Interlaced conifer branches were used to finish the remainder of the walls. The sedentary fishery undoubtedly encouraged the use of more permanent building materials, probably including the sod roofs suggested by some views and plans.

The sheltered work area thus provided was simple in layout and use.[29] The lower part of the end wall facing the water was probably left open as shown in Figure 16 to enable fishermen to slide the cod from the end of the stage to inside the shelter. A large table was set up where the splitting crews quickly dressed the fish. The dressed fish were then carried past a large low rectangular salt bin to where the *saleur* (salter) salted the fish down in piles. Each day's catch was piled separately and not mixed with that of other days. The building protected the dressed cod from the elements and allowed it to stay salted for the necessary week to 10 days. This was an important function and if not performed by a shed on the stage then a building on shore served this same purpose. Le Brun had a storehouse on shore for salting cod, as well as a stage at his Petite Brador establishment.

In spite of its utility in the loading and unloading of supplies and fish, and as a work area, not every fishing property had a stage. A detailed plan of the Côte du Nord fishing properties in 1734 illustrated only 15 stages for more than 25 fishing properties.[30] In part, this might be explained by the heavy concentration of the schooner fishery in Louisbourg harbour. With the dressing and salting of the catch done on board the schooner, the stage's functions as a work area were redundant. This left only the landing of the catch as a justification for the stage, and even this was unnecessary if small boats were used to transfer the catch ashore. Guillaume Gothier rented a Côte du Nord property in 1750 for use with a fishing schooner, even though the property apparently had no stage.[31] In the shallop fishery, dressing and salting the day's catch could be done in a storehouse near the shore, even if this was somewhat less convenient from the point of view of waste disposal. With a stage the head and entrails could simply be dumped into the water through a hole in the floor. A more simple alternative would have been to dress the fish on a table set at water's edge and to use the storehouse only in poor weather.

Surprisingly enough, neither Denys nor Du Monceau noted the practice of washing the blood and evisceral material from the dressed fish prior to salting. Without this washing, blood spots in the dried fish were unavoidable. As both writers stressed the greater value of a white finished product, it was probable that French fishermen performed this necessary task. After the fish had been salted long enough to allow the drying process to start without fear of corruption, the fish were washed of excess salt and impurities. This practice encouraged the utilization of a large open-topped lattice-work wash cage set in shallow water. The lattice work prevented the shore workers from losing any of the salted fish while washing them. Called a *timbre* by Denys and *lavoir* by Du Monceau, these wash cages are illustrated in Figures 16 and 24.[32] A wash cage was included among the items inventoried in the fishing property of Sieur Dessaudrais Robert in 1720.[33] While the excess salt was being washed off, the remaining black intestinal lining of the cod was rubbed away so that the finished product would have a nice white appearance. Ideally, the washed cod were immediately placed on wicker mats or handbarrows to prevent sand or grit from adhering to the fish.

The need to drain and press the washed fish before drying them led to the development of specialized facilities. Draining eased the drying process and pressing improved the finished appearance of the fish. The washed fish were piled on a platform that permitted drainage and kept the fish clean. At its simplest, this platform could have merely been the handbarrow on which the fish were carried from the wash cage. A *clai* or platform, built like a small fish stage, was also utilized.[34] The "une Glefs" mentioned in the Dessaudrais Robert inventory probably represented a phonetic spelling of the *clai*.[35] Denys called a similarly constructed platform a *galaire*, but described it as having an arborlike roof of branches to protect the cod from excessive heat during the summer season.[36] A simplified form of draining platform might be the explanation of the unidentified raised structures shown on the beach in front of the cabin on the plan of Lartigue's Côte du Nord fishing establishment, shown in Figure 7. Although draining platforms are not mentioned in descriptions of Isle Royale fishing properties, their simple construction and general utility certainly suggest their presence.

Three types of drying apparatus (beaches, flakes, and *rances)* were used in France's North American cod fishery.[37] Beaches were gravel beaches, either man-made or natural, which were relatively level and devoid of large rocks, sand, soil, or vegetation. Flakes were long, narrow, drying platforms made of two parallel lines of short upright poles, joined by stringers at the top with closely spaced crosspieces covered with conifer branches stripped of their vegetation. *Rances* were simply boughs and branches lying directly on the ground, in the case of non-gravelly areas, to prevent the drying fish from getting soiled by sand or dirt. All three drying devices were used in Isle Royale. For example, both the 1734 and 1742 lists of concessions on Louisbourg's Côte du Nord frequently described individual lots as being "partie en grave, partie en vigneaux."[38] La Roque presumably described *rances* when he noted in his 1752 census that the drying areas of l'Indienne and Petite Brador were made of birch and wild cherry branches rather than the stone used at

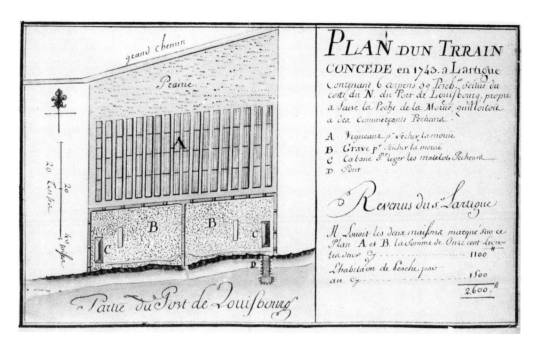

Figure 7. A detailed plan of the Lartigue fishing property of Louisbourg's north shore, seen as property 36 in Figure 5. (Source: Map Collection, Fortress of Louisbourg, No. ND-121.)

Louisbourg.[39]

Typically, to begin the drying process, the washed, salted cod were placed on the flakes after draining. As the drying surface of the flake was 2-3 feet (1 foot = 0.3048 m) above the ground, the fish were exposed to cooling breezes which helped to prevent any excessive buildup of heat. Each flake was approximately 4 or 5 feet in width, so that the fish drying in the middle could be easily reached for the repeated turnings that were necessary. In addition, each flake was separated from its neighbour by a pathway several feet wide so that the fish could be moved easily to or from the flake according to its stage in the drying process. These handling procedures resulted in long lines of narrow flakes standing parallel to each other. A plan of Louisbourg in 1720 shows this patterned arrangement of flakes at the fishing establishments at Pointe de Rochefort and near the Petit Etang.[40] An even more ordered layout of flakes appears in the plan of Lartigue's Côte du Nord fishing property shown in Figure 8.

The number of flakes required for a particular fishing establishment depended not only on the size of the season's anticipated catch but also on the amount of beach available. Early plans of Louisbourg do not show flakes

as part of the fishing establishments on the Grande Grave towards Pointe de Rochefort but do show them in areas where the beach was more restricted. Whether this illustrated a preference for use of beaches or merely reflected the difficulty of building flakes on a gravel beach is a moot point. However, George Lasson, with 107 *toises* by 13 *toises* of his fishing property devoted to beach, still had an additional 30 *toises* by 40 *toises* of flake area.[41] Certainly flakes covered sizeable areas of other shore establishments. Flakes at Le Brun's Petite Brador establishment covered an area of 50 *pieds* by 120 *pieds*, while they covered approximately 30 *toises* by 75 *toises* of Lartigue's Côte du Nord property.[42] Guillaume Gothier rented 17 flakes in 1750 for drying the catch of his summer fishery.[43] Although the width and, more importantly, the length of these flakes were not specified, their numbers alone indicate a relatively large total drying area.

Although the construction of flakes was relatively cheap in terms of materials, it was also relatively labour intensive. The trees used were smaller than those used for building stages but the gathering of the needed materials must have become more bothersome as the forests around the harbours, particularly

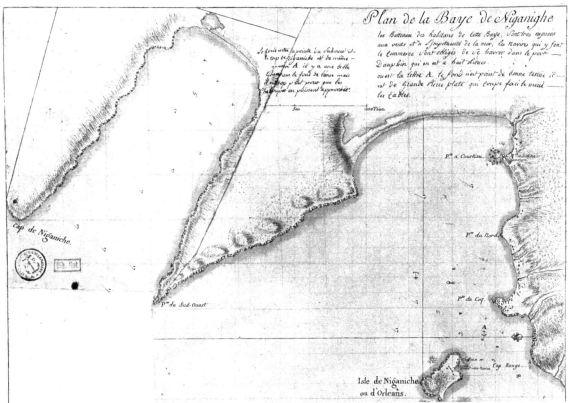

Figure 8. Plan of Ingonish Bay. As on Louisbourg's north shore, these Ingonish fishing properties followed a strip development along the shoreline. (Source: Map Collection, Fortress of Louisbourg, No. ND-130.)

Louisbourg, were cut back. The road between Louisbourg and the Lorembecs was considered to have opened up a new source of wood for flakes as well as stages. As the flakes were vulnerable to damage, attention had to be paid to their repair and upkeep. Rental agreements such as that of Guillaume Gothier specified the landlord's responsibility to provide flakes in good condition at the beginning of the fishing season.

Like flakes, beaches were a common feature of the Isle Royale fishing property and often represented a substantial investment in labour, if not capital. Some beaches, such as the Grande Grave at Louisbourg and that of Plaisance, were natural seaside gravel beaches requiring only a minimum of levelling or vegetation removal to prepare them for use. Others necessitated considerably greater effort, such as the beach prepared by Beaubassin Sylvain et Cie at their Côte du Nord property. There two soldiers preparing the land had to bring gravel to cover the large rocks.[44] Of course, using too fine a gravel

limited the circulation of air under the drying fish, so small rocks had to be used. Although beaches were used in the latter part of the drying process, they were typically found near the shoreline separating the more sequential work areas of fish stages and flakes. This positioning presumably reflected the use to which shoreline deposits of rock were put in making the beach. Recent archaeological excavation of a site in Louisbourg's Fauxbourg district revealed just such a beach in which varying sizes of rocks were used to cover the original sod layer. Interestingly enough, many of these rocks were not rounded beach stone, indicating additional sources of material were sometimes needed.[45] As these beaches would have been difficult to walk on, paths such as the one shown on Lartigue's property would have been necessary for well-travelled routes.

In addition to their initial construction, beaches required varying amounts of annual maintenance. Nicolas Denys noted, in describing the French migrant vessel fishery, that on old beaches the ship's boys had to

28

remove the grass growing amidst the gravel prior to its use.[46] Fishing proprietors would have undoubtedly ensured that their beaches were maintained in a similar fashion. Sea erosion, particularly as a result of winter storms, posed a serious threat to shoreline beaches. As a result of the damage done to the Fauxbourg by a storm on 12 January 1740 a protective sea wall was proposed and eventually built.[47] Without annual maintenance many of Isle Royale's beaches would have suffered the fate of Pierre Bonain *dit* La Chaume's obviously neglected beach at Saint Esprit. At the time of his inventory in October 1734, in addition to a stage "tres delabre," the beach was "emportez par Lamer qui ymis du Sables."[48]

As the fishing season progressed piles of dried and drying cod became a common sight on every fishing property. The ever increasing size of these piles quickly necessitated a change of drying location from the flake to the beach. The piles were carefully arranged so that fish were laid skin upwards in an overlapping fashion to retard moisture penetration. Towards the end of the drying process of a particular batch of fish, when the fish were piled in larger quantities for longer periods of time, the piles were often covered with sailcloth for added protection. As dried cod was vulnerable to water damage and Isle Royale's climate showed a marked propensity to cause such damage, more permanent shelter inside was required for the finished product. Hence, there are references to piles of cod in cabins and storehouses on the fishing properties and in storehouses in Louisbourg.[49]

In addition to these specialized work areas for the processing of the fish, the typical fishing property contained several buildings that also served specialized functions. Generally speaking, these buildings would have included the house of the fishing proprietor, cabins for the accommodation of the fishing crews, storehouses for the storage of fishing and food supplies, and assorted outbuildings. As one might expect, the size, type of construction, and quality of these buildings varied according to the needs and means of the individual fishing proprietor.

Although not every fishing property served as the dwelling place of its owner, the house of the fishing proprietor was a prominent feature of such properties. In addition to providing living quarters for the fishing proprietor and his family, part of the house served as his office. Possibly consisting only of a table or desk, it was here that the fishing proprietor kept his important papers and accounts. By law he was required to keep written contracts with his crew, a register of the purchases and sales of his fishing operation, and a register of his business transactions with his men.[50] Common business sense and possible court action ensured that he also kept his bills and receipts, the titles to any land he owned, and any rental agreements he might have made. Inventories, such as that of Elie Tesson La Floury, indicated that part of the house also acted as storage space for small items that might have otherwise "disappeared."[51]

The hired fishermen always appear to have been housed in cabins separate from the house. Little information is given concerning the furnishings or cooking and heating arrangements of these buildings. There is some evidence, however, that the arrangements were spartan. The assessors of François Blondel's estate were told by the fishermen in the *cabanne des pescheurs* that the only item there belonging to Blondel was a copper *chaudière* for making soup.[52] So many fishermen's inventories contain straw ticks that it seems likely the cabins were equipped with rows of bed frames. These may have followed the simple double bed described by Denys, which had a wooden frame and an interwoven rope bottom with a centre pole to keep the two men from sleeping in a heap in the middle.[53] Testimony in a criminal process against a *maître de grave* or shoremaster in Laurembec indicated that each fisherman had his own bed.[54]

Storehouses formed an integral part of every fishing property in Isle Royale. Whether they were attached to the fishing proprietor's house or were detached buildings in their own right, storehouses provided the necessary storage space for the large volume of supplies required in the sedentary fishery. Salt, commonly used at the rate of slightly more than 1 *barrique* (hogshead) per 10 *quintaux* of cod, was but one of the needed fishing supplies.[55] Fishing equipment, such as hooks and lines, and replacement gear for shallops and schooners were also kept on hand in varying quantities. Food for both the fishing proprietor's family and for the fishing crews comprised another large commodity group. Storehouses appear to have been frequently equipped with storage bins to keep biscuit. In addition, the fishing proprietor would keep on hand a large number of items for resale to his crews. Liquor would have been evident in abundance, as well as food items to supplement the basic diet and articles of clothing.

There were also specialized outbuildings, the number and type of which varied according to the degree of self-sufficiency the fishing proprietor desired to attain. Whereas Bonnaventure Le Brun had an oven on his Petite Brador property, the fishing property of the widow Perré was supplied with bread from a baker.[56] Similarly Le Brun's Niganiche property had facilities for his carpenter, smith, and cooper-tradesmen, not commonly found on other fishing properties. Most fishing properties had a large cooking pot for brewing *sapinette* or spruce beer. Sheds for livestock and gardens were also relatively common as fishing proprietors supplemented food purchases with home-grown products. In addition to the more typical three cows and a calf, Elie Tesson La Floury also kept two horses for use with his cart.[57]

The typical fishing property at Louisbourg and elsewhere in Isle Royale represented a fairly substantial investment in capital and labour. A necessary adjunct of the dry fishery, the shore property necessitated the development of relatively specialized work areas and buildings. Its cost, both in terms of natural factors and economic investment, restricted access to the dry fishery. However, the relatively small size of the typical fishing property mitigated against the assemblage of large numbers of shore properties and the creation of monopoly conditions. Indeed, the colony's short history and, hence, truncated economic development ensured that even at the height of the Isle Royale fishery production the fishing proprietors were unable to use all the available shoreline property.

METHODOLOGY OF THE ISLE ROYALE
FISHERY

As might be expected, the dry fishery, whether conducted by residents or migrants, dominated the fish production of Isle Royale during both French occupations. The methodology of France's extensive green fishery ensured that those vessels had only minimal contact with the colony. Of course, green fish were produced at Isle Royale towards the end of the fishing season when it was too late to begin drying, but this amount was small. For example, a 1737 list of Isle Royale's exports, excluding those to New England and Acadia, contained dried cod to the value of 1 389 380 *livres,* but only 6736.5 *livres* worth of green cod.[1] The bulk of this latter amount (worth 6000 *livres*) was taken by a single ship from St. Jean de Luz which had arrived at Niganiche on 21 October 1737 for its second trading voyage of the season.[2] Dried cod similarly dwarfed the export values ascribed to cod by-products such as salted eggs, sounds (air bladders), and tongues, and to a lesser extent those of cod oil.

The dry fishery of Isle Royale utilized methods of fish catching and processing that European fishermen had used on the coasts of North America for generations. More particularly these fishermen drew on the experience of French migrant and resident fishermen at Plaisance and the Petit Nord in Newfoundland, at Gaspé, and on the coasts of Acadia. At this time, fishing was done through the simple expedient of handlining in which each fisherman used one or two lines with one or two hooks on the end. Long lines and trap nets were still devices of the future. Processing was through drying, in which the dressed and salted cod underwent a long process of repeated exposure to sun and air until the desired degree of dryness was achieved. With none of the modern aids, such as artificial dryers, fishermen had to contend with the vagaries of the weather while drying their catches.

Although Isle Royale fishermen utilized the basic methodology of the dry fishery, some adaptations had to be made to local conditions. The principal difference between Isle Royale and Plaisance was the fishing grounds to be exploited. The cod stocks of the immediate inshore waters around Isle Royale were less abundant than those of Plaisance. Basque merchants complained in 1716 that the fishing grounds were "plus de six lieues" from shore.[3] With such a distance to travel the fishermen were not always able to return to shore that day and occasionally fishing crews had to spend a dangerous night at sea in their small shallops. The Basques recommended a territorial exchange, with France giving Isle Royale to Britain in exchange for Newfoundland. Although the Basques advanced a number of reasons why the British would acquiesce to this proposal, such an exchange was unlikely given the British determination to obtain Plaisance by the Treaty of Utrecht.

As the likelihood of a territorial exchange remained remote, French fishermen adapted their technology to meet existing conditions. An obvious solution was to employ larger, more seaworthy vessels to enable fishermen to exploit the offshore fishing grounds, but this necessitated other procedural changes. The use of larger vessels encouraged fishermen to make longer trips to the fishing banks, to make more efficient use of the vessels' greater carrying capacity, and to obtain a more favourable ratio between the time spent fishing and that spent in transit to and from the banks. The longer trips necessitated dressing and salting the catch be done on board the fishing vessel rather than on the stage as in the inshore fishery. In addition, the longer period of time between dressing and drying required these offshore fish to be more heavily salted, resulting in a slightly poorer cure. However, the greater productivity of the vessel fishery was expected to compensate for the lower value of the finished product.

Isle Royale fishermen quicky expanded their operations to include the offshore banks but this move was not made without controversy. As early as 1716 Isle Royale fishermen were reported fishing "en bateau" at a distance of 4 or 5 leagues from shore.[4] The dumping of fish offal overboard from these vessels was reported to be keeping the fish offshore, causing poor catches for the shallop fishery. An attempt to have this offshore fishery prohibited failed to gain support from either resident or migrant fishermen and members of both groups adopted the new technology.

Charroys and *brigantins* were the initial

Figure 9. Ships used in the dry cod fishery. The upper vessel is called a *flivot*, and the lower one is a vessel of about *100 tonneaux* from St. Malo. (Source: Du Monceau, *Traité générale des pesches . .*, Part II, Section 1.)

Figure 10. Cod fishing on the Grand Banks is carried out using ships of different sizes and types. The men are seen fishing along the side of the vessels protected by wind screens. In this instance the catch would be preserved only through salting and not drying. (Source: Du Monceau, *Traité général des pesches...*, Part II, Section 1.)

rigs used in the offshore fishery[5] but these were soon discarded in favour of the schooner fore-and-aft rig. This rig proved extremely handy for navigating inshore as these fishing vessels were frequently required to do. Although schooners were owned in the colony since its founding, they did not appear in the colony's fishing statistics until 1721. The French were already familiar with New England's use of fishing schooners off the coast of Acadia. Indeed, the Basques had referred to this New England employment of schooners to support their allegations of British interest in exchanging Newfoundland for Cape Breton.

The lack of per unit production figures

Figure 11. Cross section of a Norman vessel practising the green fishery on the bank. Once again the fishermen line the ship's side, a dressing crew in the centre cleans and splits the catch, and a lone salter salts it down in the hold. (Source: Du Monceau, *Traité général des pesches ...*, Part II, Section 1.)

prohibits an evaluation of the total catch landed by schooners during the early years, but thereafter schooners accounted for a significant proportion of the colony's total dried fish production. As shown in Table 6,[6] schooners typically accounted for under a quarter of the colony's total dried fish production during the mid-1720s and early 1730s, for just over a quarter through most of the mid-1730s, and for over a third in the late 1730s and early 1740s. The outbreak of war in 1744 witnessed a drop in the schooners production back to a quarter of Isle Royale's total fish production. Missing data, particularly in the 1720s and the early 1740s, makes these trends somewhat conjectural.

The most striking contrast between the two branches of the fishery was the actual physical difference between the shallop and the schooner. Indeed, it was this physical contrast that formed the basis for all the other differences between the two fisheries. The shallop was simply a small undecked rowing boat equipped with a small mast that could be stepped or unstepped as necessity dictated. The schooner, on the other hand, was a full-decked vessel capable of a protract-

ed stay on the offshore banks. By definition the schooner had two masts with fore-and-aft rigging on each mast. Du Monceau described shallops as having a carrying capacity of 4 or 5 *tonneaux*.[7] *Commissaire-ordonnateur* Prévost noted that schooners typically ranged in size up to 60 *tonneaux* and that those under 30 *tonneaux* were impractical in size.[8]

Although no detailed descriptions of the shallop's construction in Isle Royale have as yet been found, descriptions by Denys and Du Monceau, combined with pictorial evidence of contemporary plans and views, give a fairly complete picture.[9] As shown in Figure 16, the shallop tapered from the middle towards each end and was easily identified by its pronounced stem, stern, and keel, and by its full rounded bows. They also reached a respectable size and seaworthiness. Louis Gilbert was contracted in 1750 to build several fishing shallops of 28-29 *pieds* of keel length.[10] Voyages from Louisbourg to Niganiche and even to Newfoundland were not unusual. Traditionally the shallops were prepared for fishing by dividing the interior into six compartments.[11] This was accomplished by nailing appropriate-sized boards or barrel

Table 6. Productivity of the *chaloupe* and *goélette* fisheries of Isle Royale

Year	Chaloupe		Goélette		Total
	quintaux	% of total	*quintaux*	% of total	*quintaux*
1715					
1716					
1717					
1718	156 500	100.0			156 500
1719					156 520
1720					
1721					125 600
1722					
1723	84 600	69.8	36 800	30.2	121 160
1724					
1725					
1726	113 700	80.7	27 200	19.3	140 900
1727	87 880	76.6	26 800	23.4	114 680
1728					
1729					
1730	128 030	77.3	37 600	22.7	165 530
1731	137 940	82.3	29 600	17.7	167 540
1732					
1733	117 685	71.1	47 500	28.9	165 530
1734	98 210	70.2	41 600	29.8	139 810
1735	104 535	73.4	37 960	26.6	142 495
1736	117 180	77.5	35 960	22.5	151 110
1737	117 500	78.7	31 800	21.3	149 300
1738	112 820	74.0	39 650	26.0	152 470
1739	92 660	64.5	51 000	35.5	143 660
1740	78 300	69.6	44 850	36.4	123 150
1741					
1742					83 410
1743	57 520	64.8	31 200	35.2	88 720
1744	52 430	75.5	17 000	24.5	69 430
1750					90 560
1751					95 580
1752					83 130
1753	62 450	63.4	36 000	36.6	98 450
1754					
1755					

staves from the thwarts to the floor boards, giving each of the three fishermen manning the shallop a compartment to stand in while fishing and another in which to stow his catch. Denys stated that the shallops were also equipped with canvas sideboards. Other gear included a square or lateen sail, three oars, a grapnel, a compass, fishing tackle and bait, a small barrel of watered wine or beer, and a basket of biscuit.

The schooner was a considerably more substantial vessel than the shallop. Although detailed measurements for the schooner are also scarce, those in existence indicate that a schooner of *50 tonneaux* was some *50 pieds* in length.[12] Like the shallop, the schooner had a pronounced stem, stern, and keel, and similar full rounded bows. The bowsprit was

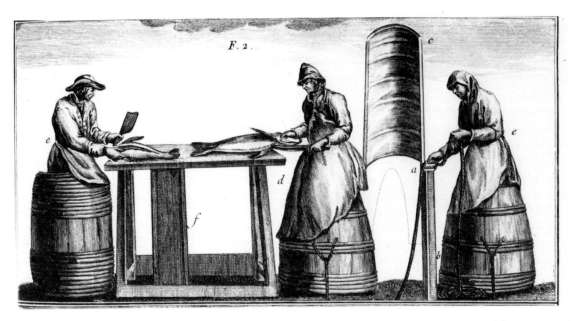

Figure 12. Detail of the splitting table on board a fishing vessel, with the throater (d) at one end and the splitter (e) at the other; a fisherman (e) uses his handline. The barrels were used to protect the fishermen from the water and blood and for stability. (Source: Du Monceau, *Traité général des pesches ...*, Part II, Section 1.)

Figure 13. The discharging of salted fish which are divided into lots according to their size and loaded onto wheelbarrows for transportation to the storehouses. (Source: Du Monceau, *Traité général des pesches ...*, Part II, Section 1.)

abnormally long by modern standards and there was a raised quarterdeck. It is important to note that not all vessels thus employed in the bank fishery were necessarily schooner rigged. Censuses recorded the numbers of *batteaux* and schooners employed in the fishery.[13] *Batteaux* presumably referred to a variety of riggings on one- and two-masted, decked vessels.

Although there are indications that the size of the shallop itself at Isle Royale may have increased compared with its counterpart at Plaisance, the fishing crew manning it remained fixed at three. Denys distinguished these men as the *maître* (master), *beaussoin* (bosun), and the *arimier* (stower).[14] The master was typically the most experienced and was in charge of steering the shallop and in choosing the particular fishing ground. The bosun was next in experience and on the pay scale; he sat in the bow of the shallop and took charge of anchoring the shallop on the fishing ground and in guiding its approach to the stage. The stower was the least experienced and his only specialized duty appears to have been to stow the catch properly to keep the shallop in trim. In 1715 St. Ovide confirmed the bosun's pay as higher than that of the stower but this distinction quickly fell into disuse.[15] Thereafter, the bosun and stower were referred to interchangeably as *compagnon-pêcheurs* (companion fishermen) and the only pay distinction was between them and the *maître de chaloupe* (shallop master).[16] In addition to the actual fishing crews two shallops required a shore crew of four men. This crew included the *maître de grave* (shore-master), the *saleur* (salter), a *décoleur* (header), and a *garçon* (boy). This crew undertook the multitude of tasks associated with drying the cod on shore. The shoremaster oversaw the drying process and ensured that the proper procedures were done on time. The salter formed part of the dressing crew, and the boy peformed the odd jobs associated with a fishing establishment. In some instances another boy was hired in place of the header.

The greater size of the schooner compared with that of the shallop demanded an increase in the number of crew employed. The act of 1743 regulating the fishery specified a crew of 11 — a *maître* (master), a header, six companion fishermen, a salter, a *trancheur* (splitter), and a boy.[17] This probably represented a maximum crew, as a schooner's production necessitated a shore crew similar in size to that for two shallops. A memoir of 1733 noted that Basque fishermen were fishing from schooners with crews of six or seven men.[18] In this instance, the schooner crew probably consisted of the master and the companion fishermen, with one of the fishermen also acting as the salter. The remainder of the 11-man crew would serve on shore with a division of functions similar to that of the shallop fishery.

On board both shallops and schooners fishing was done with hook and line. Each fisherman was equipped with two lines and if the fish were plentiful each line might be equipped with two hooks instead of one. As cod are bottom feeding fish each line had a lead weight of several pounds as a sinker. In some instances the hook formed an integral part of the sinker which was shaped like a small fish to act as a lure. Otherwise the hook had to be baited.[19] (Several of these hooks and lines are illustrated in Figures 23 and 25). At Isle Royale, mackerel and herring were the most common bait and it was an additional duty of the shore crews to set nets for catching these bait fish, as shown in Figure 18.[20] In addition, the fishermen in the shallop were equipped with mackerel lines so that they could catch fresh bait if the opportunity presented itself.[21] As the schooners were on the fishing banks for extended periods of time, the acquisition of fresh supplies of bait was frequently impossible, so salted bait had to be used. However, salted bait was not considered as good as fresh bait. In the absence of other supplies of bait the hook could always be baited with offal from a freshly caught cod. In the inshore fishery the hooks, lines, and sinkers were generally lighter than those used on the banks.

An important difference between the schooner and shallop fisheries came in the area of per unit production. The shallop fishery was broken into a summer and a winter season, while the schooner fishery was restricted to the summer. In 1739 *commissaire-ordonnateur* Le Normant estimated the average per unit production of these three branches as 300, 120, and 840 *quintaux*, respectively.[22] It is interesting to note that the highest per unit production figures in the annual statistical returns also occurred in 1739. Except for the winter fishery Le Normant's estimates appeared high when compared with those taken from these returns and shown in Table 7.[23] During the 1720s and 1730s there was a gradual increase in per unit production in the summer shallop

Table 7. Average per unit production in Isle Royale shallop and schooner fisheries, 1715-55

Year	Shallop		Schooner
	Summer qtx./unit	Winter qtx./unit	Summer qtx./unit
1715			
1716			
1717			
1718	250		
1719	250		
1720			
1721			
1722			
1723	180	100	400
1724			
1725			
1726	210	130	400
1727	200	110	400
1728			
1729			
1730	210	130	400
1731	220	110	400
1732			
1733	215	100	500
1734	230	80	520
1735	245	110	520
1736	250	110	580
1737	250	100	600
1738	250	120	650
1739	260	180	850
1740	200	90	650
1741			
1742	150		600
1743	200	40	600
1744	200	30	500
1750			
1751			
1752	260	35	750
1753	230	45	720
1754	250		760
1755			

and the schooner fishery. The winter shallop fishery, with the exception of a few poor seasons, fluctuated between 110 and 130 *quintaux* per shallop. After the 1739 season there was a decrease in per unit production in all branches, which reflects in part outfitting problems due to wartime tensions and also British interference with the shallop fishery on Isle Royale's southeast coast.[24] The extreme drop in the per unit production of the winter shallop fishery appears to have been due to a change in fish migratory patterns.[25] This situation does not appear to have changed by the early 1750s.

The daily routine of the fishermen formed one of the most striking contrasts between the shallop and schooner fisheries. Generally speaking shallop fishermen were closely tied to the shore establishment, leaving for adjacent inshore fishing grounds early in the morning. They would spend the day fishing on these grounds at distances up to 4 or 5 leagues from shore and return home late in the afternoon.[26] While the fishermen were out fishing the shore crew would be busy washing and stacking any fish taken out of salt, spreading and turning the drying fish on the flakes and beaches, and piling the dried fish if required. When the shallop returned to the stage, both fishermen and shore crew joined in dressing and salting the catch. This routine was repeated daily as long as both the weather and the availability of fish permitted. Sundays provided a welcome day of rest in both fisheries, enabling fishermen to wash or mend their clothes and equipment, to relax, and to attend divine services if the opportunity presented itself.

Although the shallop fishery was conducted inshore and on a daily basis, it was not without danger to the fishermen. To maximize catches fishermen had to sometimes go in questionable weather, and sudden wind squalls and quickening gales frequently resulted in loss of life. One fisherman who survived such a fate was LeBreton of Scatary. In 1735 he reported that he had been fishing in his shallop one year when a rising wind forced him 12 or 13 leagues off shore.[27] LeBreton was fortunate and survived this storm but other fishermen died in similar circumstances. The winter shallop fishery was even more dangerous. On 10 December 1734, Mathieu Laisne, Noel L'Eclanche, and Joannis L'Etapy, three fishermen in the employ of François Lessenne, drowned in the harbour of St. Esprit where Lessenne had a fishing establishment.[28] Individuals entering the fishery recognized these dangers and some set their affairs in order as a precautionary measure. Jacques Germain, an armourer who was entering the summer shallop fishery at Lorembec, had his will drawn up on 22 April 1743.[29]

The fishing schooners pushed considerably

Figure 14. The preparation of stockfish in northern Europe. The cleaned fish are washed (A), piled to drain (B), and allowed to dry on beaches (C) or on racks (D). (Source: Du Monceau, *Traité général des pesches ...,* Part II, Section 1.)

Figure 15. Drying cod on rocks and large stones; the practice of leaving the heads on was not followed in North America. (Source: Du Monceau, *Traité général des pesches ...,* Part II, Section 1.)

Figure 16. This plate illustrates almost all the objects and procedures used in the preparation of dried cod: a shallop unloading fish at the stage (A); the cabin protecting the headers, splitters, and salter (C and D); cod drying on flakes (F) and on the beach (H); a wash cage (K); and dried cod in piles (L). (Source: Du Monceau, *Traité général des pesches ...*, Part II, Section 1.)

further afield than did the shallops. Schooners operating out of Isle Royale fished on the banks on the Scotian shelf from Sable Island northwards. They also fished on St. Pierre bank off the south coast of Newfoundland.[30] If these fishing grounds proved less successful, as was frequently the case later in the season, then the schooners shifted their area of activity to the banks in the Gulf of St. Lawrence.[31] The length of time a fishing schooner remained out on a trip varied according to the size of the vessel and the success it enjoyed fishing, but trips averaged between 20 and 40 days.[32] As in the inshore fishery, the men fished from morning until late afternoon, conditions permitting, and then dressed and salted their fish. Fishing was done with handlines from the deck of the vessel and not from smaller boats at a distance from the vessel.

Although schooners were more seaworthy than shallops and did not engage in the dangerous winter fishery, schooner fishermen nevertheless encountered risks while plying their trade. In December 1742 the officers of the Admiralty at Louisbourg reported that two fishing vessels had been lost at sea by being rammed by other vessels.[33] Fortunately the other vessels succeeded in rescuing the crews of the stricken vessels. To avoid other such accidents the Admiralty officers recommended that vessels be required to carry a light at the top of their masts at night and that at least two men be on watch. Furthermore the officers suggested that any vessel rammed which had not shown a light be the only one required to pay damages.

Both fisheries also differed somewhat in the procedures for processing the fish prior to drying. In the shallop fishery the fish were dressed and salted on the staging after the shallop returned to shore in the afternoon. As drying could generally be started immediately after the salt had penetrated the fish, the day's catch only had to be lightly salted for

Figure 17. Detail of the stage where the fish were cleaned, split, and salted. The large low box in the background contained salt. (Source: Du Monceau, *Traité général des pesches ...*, Part II, Section 1.)

several days. In contrast, fish caught on board a schooner were dressed and salted down directly on board the vessel. Because a fishing voyage lasted several weeks or longer, the fish had to be heavily salted to preserve them until they could be landed for the drying process. This heavier salting meant these fish could not be dried as hard as those produced by the shallop fishery and consequently were considered less valuable.

Either a three- or two-man splitting crew was used to dress the fish.[34] With a three-man crew the functions were divided into those of the *picqueur* (throater), header, and splitter. All three men stood at a splitting table and performed their duties in sequence. First the throater, armed with a knife, took a cod and cut its throat horizontally just below the gills and then gave it a second cut lengthwise along the stomach from the first cut to the anal opening. The header then took the fish and removed the viscera, throwing the roe and the liver into separate baskets if they were to be saved. Next the header broke off the cod's head with his hands and either dropped it into the water or threw it to one side if someone had been detailed to remove the tongues. The splitter then removed the

backbone and ribs from the anal opening forward so that the fish lay flat. The *noues* or air bladders of the cod could also be removed from the backbone. The dressed cod were then carried to where the salter carefully salted them down in piles, making certain that each day's catch was salted separately and was not intermixed with those of previous catches. Although contemporary accounts neglect to mention it, the dressed fish were probably washed before salting. Failure to wash the fish at this time would have left the fish discoloured and marked with blood spots.

When the fish were taken from salt to begin the drying process they were first washed in seawater. This removed any blood or visceral material on the cod as well as any excess surface salt. It was at this time that any of the remaining black abdominal lining could be rubbed off. Its removal did not affect the drying process but merely improved the appearance of the finished product. After washing the fish were stacked until the excess water had drained off before being laid out to dry. Although the fish could be laid out on gravel beaches, on branches lying on the ground, or even on large rocks, at Isle Royale the fish were typically placed on flakes to dry.

Fig. 1.

Figure 18. Fishing for capelin for use as bait in the cod fishery. Nets were also used at Isle Royale for catching herring and mackerel as bait. (Source: Du Monceau, *Traité général des pesches ...*, Part II, Section 1.)

They were initially placed skin side up but were thereafter placed flesh up. In the evening the fish had to be turned skin up to protect the flesh from the dew, and every morning they were again turned flesh up. In the event of fog or rain the fish had to be turned skin up or covered up or even taken inside. During hot weather the fish had to be turned repeatedly to prevent them from becoming burnt.

After several days drying on the flakes an extended piling process was begun. When the fish were turned over at night they were formed into small piles, using as a base two fish lying side by side and head to tail. Initially these piles consisted of about half a dozen fish but they were gradually increased in size until each pile contained about 25 fish. At this point the pile of fish was transferred to the beach. Here the piling process was continued, with the fish being formed in increasingly larger piles and then being spread to dry on the beach between pilings. Initially the cod were kept in piles only overnight but as the piles increased in size the fish were kept stacked longer, with the largest piles remaining stacked for several days. Piling the fish helped protect them from moisture damage as they became drier and also enabled the moisture in the centre of the fish to work its way to the surface where it could evaporate. When the fish were finally considered to be dry enough they were stacked in large piles. The fish were carefully arranged so that

Figure 19. Front and back view of a round cod (E and F) and the same view of a split cod from which two-thirds of the backbone has been removed (G and H). (Source: Du Monceau, *Traité général des pesches ...*, Part II, Section 1.)

Figure 20. Culling or grading the fish according to size, quality, and species. In the green fishery, culling would ordinarily have taken place on the deck of the vessel. (Source: Du Monceau, *Traité général des pesches ...*, Part II, Section 1.)

Figure 21. Labour specialization in the fishery: A, a splitter holding his knife; B, a shoreworker lifting a cod with a pew; C, a shoreworker putting salt on a small shovel. (Source: Du Monceau, *Traité général des pesches ...*, Part II, Section 1.)

Figure 22. Labour specialization in the fishery: A, salter salting the catch; B, shoreworkers bringing salt to salter. In the shallop fishery the central figure would represent the shoremaster directing activities. (Source: Du Monceau, *Traité général des pesches ...*, Part II, Section 1.)

moisture could not enter the pile and the whole pile might be covered with sailcloth for further protection. If the dried cod were to be stored any length of time it was preferable to place them in a dry storehouse.

Bad weather could hinder both the catching

Pl. VII.

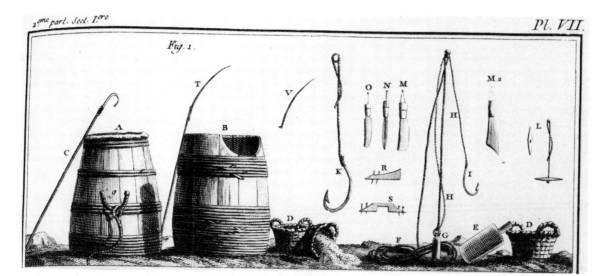

Figure 23. Utensils used in the Grand Bank fishery: A, barrel in which fishermen stood to keep dry; B, barrel for holding livers; C, a gaff; D, baskets for carrying salt, livers, tongues; E, small shovel used for salting; F, fishing line; G, lead sinker; I and K, fish hooks; L, instrument used in cutting out cod tongues; M, N, and O, heading and splitting knives; Ma, knife used for removing the sounds; R and S, braces for securing barrels to the vessel's deck; T, a pew for lifting cod; V, the tine of a pew. (Source: Du Monceau, *Traité général des pesches ...*, Part II, Section 1.)

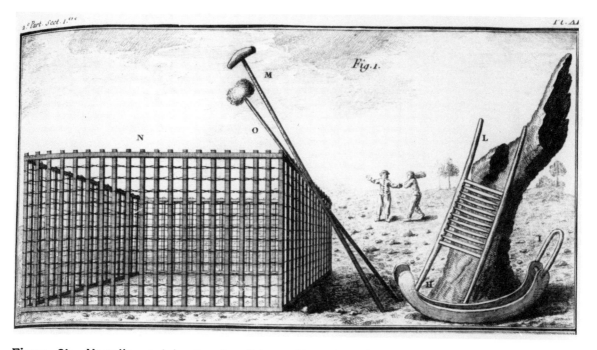

Figure 24. Utensils used in the dry fishery: H, a drag for moving fish on the stage; L, a handbarrow for carrying fish; M and O, utensils for washing salted cod; N, a wash cage for holding salted cod during washing. (Source: Du Monceau, *Traité général des pesches ...*, Part II, Section 1.)

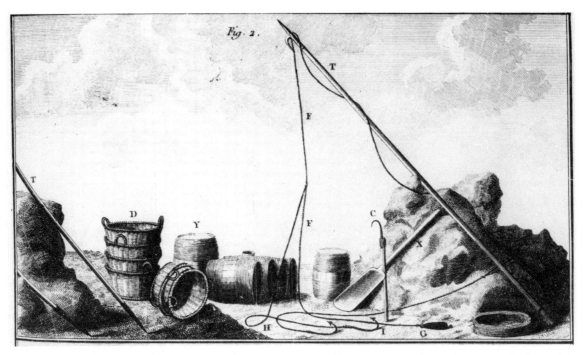

Figure 25. Utensils used in the cod fishery: C, a gaff; D, baskets for salt, etc.; F, G, and I, fishing line with lead sinker and two hooks; T, boat hooks; X, large shovel for moving salt; Y, barrels for salting tongues, etc. (Source: Du Monceau, *Traité général des pesches ...*, Part II, Section 1.)

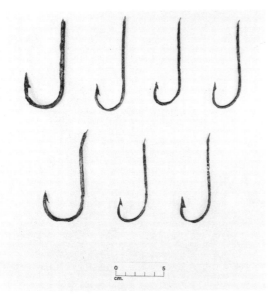

Figure 26. Selection of fish hooks excavated from harbourfront properties at the Fortress of Louisbourg National Historic Park.

Figure 27. Cod jigging reel excavated at the Fortress of Louisbourg National Historic Park.

and the drying of the fish. Storms not only stopped crews from fishing but also caused extensive damage to equipment. Two storms in November of 1726 caused so much damage to property that the advantages of a good summer fishery were lost.[35] Similarly, excessive heat or lengthy periods of wet weather proved damaging to dry cod. Port Dauphin was considered a poor fishing station due to the excessive heat caused by mountains cutting off cooling breezes.[36] Although counter measures such as turning and piling the fish could be taken for short bouts of wet weather, a lengthy stay of poor weather spelled disaster. In 1733 protracted poor weather adversely affected the drying cod and caused a considerable loss to the colony's commerce.[37] Even the most experienced fishing proprietor could not always minimize the impact and vagaries in the weather.

Similarly, changes in the accustomed pattern of fish migrations caused fluctuations in fish production, particularly in the inshore fishery. Unlike the schooner fishery, the shallops exploiting the inshore fish stocks were restricted in the distance they could go from the shore establishment. If the cod bypassed an area for some reason the shallop fishermen would have a poor season. A successful fishery in 1714 was followed by two poor seasons. Contemporaries blamed this seeming exhaustion of the fish stocks on the great increase in the number of shallops.[38] However, this shortage of fish proved to be a temporary phenomenon and was probably due to a change in migratory patterns rather than a real decline in fish stocks.

If the shortage of fish was purely local, one solution was to go fishing en degrat.[39] In this instance a temporary fishing establishment would be made at some other point along the coast where fish were abundant. The fish caught would be dressed and salted at the new station and then trans-shipped to the permanent establishment for drying. This transfer of the fishing effort was called a petit degrat with regard to shallops and a grand degrat when referring to schooners. Obviously the formation of a shore establishment applied only to those fishing from shallops. The best example of the petit degrat at Isle Royale occurred almost annually in the winter shallop fishery. During the months of November to January the cod again appeared in the colony's inshore waters but were more abundant at Niganiche and other parts of the northeast coast than in the Louisbourg-Scatary district.[40] Although only residents engaged in this fishery there was a sizeable transfer of shallops towards the northeast. Petit Brador, which had little or no summer fishery, always harboured shallops during the winter fishery.[41] An exception to this rule occurred in September 1735 when the fishermen of Scatary were at Petit Brador fishing with their shallops en degrat.[42]

The migration patterns of the cod gave seasonal characteristics to the cod fishery at Isle Royale. In the inshore waters the summer shallop fishery lasted from the beginning of May to September. There was then a break in this fishery until November, when the fish reappeared until the end of January or even into February.[43] The season for the schooner

Table 8. Annual distribution of the shallop fishery in Isle Royale for selected years 1718-53 in real numbers and as percentages

Harbour	1718 No.	1718 %	1723 No.	1723 %	1727 No.	1727 %	1731 No.	1731 %	1735 No.	1735 %	1739 No.	1739 %	1753 No.	1753 %
Daspé			17	4.8	24	6.2	14	3.6			21	7.5		
Niganiche	14	2.2	63	17.9	69	17.9	70	17.9	34	11.1	51	18.1	42	16.8
L'Indienne	6	0.9	35	9.4	15	3.9	8	2.0	4	1.3	11	3.9	5	2.0
Scatary	117	18.7	20	5.7	25	6.5	23	5.9	17	5.6	13	4.6	20	8.0
La Baleine	74	11.8	21	6.0	25	6.5	44	11.3	34	11.1	30	10.7	18	7.2
Petit Lorembec			13	3.7	23	6.0	45	11.5	36	11.8	27	9.6	54	21.6
Louisbourg	145	23.2	45	12.8	50	13.0	36	8.2	34	11.1	21	7.5	28	11.2
Fourché			7	2.0	19	4.9	31	7.9	33	10.8	23	8.2		
Saint Esprit	16	2.6	55	15.6	43	11.2	46	11.8	28	9.2	27	9.6		
Isles Michaux					6	1.6	6	1.5	4	1.3	6	2.1	9	3.6
Petit Degrat	43	6.9	24	6.8	33	8.6	26	6.6	24	7.9	26	9.3	24	9.6
Isle St. Jean					38	9.9	40	10.2	30	9.8	22	7.8	25	10.0
Isle Magdelaine							2	0.5	2	0.7	1	0.4	2	0.8
Elsewhere	184	29.4	52	14.8	15	3.9			25	8.2	2	0.7	23	9.2
Total	626		352		385		391		305		281		250	

47

Table 9. Annual distribution of the schooner fishery in Isle Royale for selected years 1718-53 in real numbers and as percentages

Harbour	1718 No.	1718 %	1723 No.	1723 %	1727 No.	1727 %	1731 No.	1731 %	1735 No.	1735 %	1739 No.	1739 %	1753 No.	1753 %
Daspé														
Niganiche			7	7.5	10	15.0	10	13.5	3	4.1	2	3.3		
L'Indienne														
Scatary			16	17.2	4	6.0	8	10.8	6	8.2	2	3.3	1	2.0
La Baleine											1	1.7		
Petit Lorembec			3	3.2			1	1.4						
Louisbourg			57	61.3	49	73.1	53	71.6	60	82.2	54	90.0	48	96.0
Fourché									1	1.4	1	1.7		
Saint Esprit			4	4.3										
Isle Michaux														
Petit Degrat			3	3.2	1	1.5			1	1.4				
Isle St. Jean									1	1.4				
Isle Magdelaine							1	1.4	1	1.4			1	2.0
Elsewhere			3	3.2	2	3.0	1	1.4						
Total			93		67		74		73		60		50	

fishery coincided with that of the summer shallop fishery but this encompassed a change in the fishing banks utilized. During the first half of the season the schooners fished on the banks on the Scotian shelf off the coasts of Isle Royale and Acadia. After about 15 July the schooners transferred their operations to the banks of the Gulf of St. Lawrence.[44] Of course, as with so many aspects of the fishery, there were no conclusive rules regarding this, and invididual preference modified many general practices.

As shown in Tables 8[45] and 9[46] the shallop and schooner fisheries were not distributed uniformly around the colony. In the shallop fishery there were three main areas of resource exploitation. First there was the Louisbourg-Scatary district, of which L'Indienne should perhaps be considered a satellite and which consistently produced from 30 to 40% of the colony's total. Second there was the southeast coast from Fourchu to Petit Degrat producing approximately 26-28% of the colony's total from the late 1720s to the 1730s. Third there was the northeast coast which had most of its production centred at Ingonish and which fluctuated around 23% of the total. After the late 1720s Isle St. Jean produced approximately 10% of the colony's total, while Isles de la Magdelaine contributed a negligible quantity during French occupations. As previously noted, there was a shift of production factors towards the northeast during the winter fishery which increased the

importance of that region in the overall production of the shallop fishery.

The schooner fishery, as shown in Table 9, suffered from a considerably more uneven geographic distribution than did the shallop fishery. The three centres of Louisbourg, Niganiche, and Scatary consistently accounted for over 90% of the colony's schooners after the mid-1720s, with Louisbourg commanding an ever-increasing share. After the first few years the southeast coast accounted for only a marginal proportion, as did Isle St. Jean and Isles de la Magdelaine. Whereas the distribution of the shallop fishery probably reflected fairly accurately the distribution of the fish stocks exploitable by contemporary methodology, this was not the case for the schooner fishery. These vessels were obviously able to land their catch for drying at any reasonable harbour. The concentration of the schooner fishery at Louisbourg probably reflected this fishery's concentration in the hands of migrant fishermen from France. As these fishermen frequently combined trading with fishing they would have been attracted by the brighter business prospects of Louisbourg as well as by the greater amenities offered by the capital. In addition, it was the practice of these fishermen to winter their schooners in the colony, and their own trading vessels would need a safe anchorage for the summer fishing season. Louisbourg's large sheltered harbour could meet both requirements and the presence of protective authority added to the attraction.

As dried fish production was labour intensive the cod fishery placed great demands on the human resources of Isle Royale. Indeed the manpower demands of the colony's total dried fish production exceeded the human resources available in the colony for this employment. The solution to this seeming labour crisis was found in the existence of a French migrant fishery that made an annual appearance for the summer fishing season. This migrant fishery supplemented the production of Isle Royale's resident fishermen, thereby greatly boosting the colony's total production. Moreover, the resident fishery was highly dependent on a pool of transient labour from France to attain its own production levels. Less than half of this transient labour force wintered in the colony; most of them simply followed the seasonal pattern of the French migrant fishery.

In spite of the colony's dependence on migrant French labour the resident fishery quickly asserted its supremacy over the migrant fishery in terms of total dried fish production. Although the official fishery statistics should not be taken as an exact account of any individual year for reasons previously mentioned, they nevertheless provide a reasonably accurate picture for comparative purposes. As shown in Table 10,[1] the resident fishery moved from approximate parity with the migrant fishery in 1718-19 to averaging slightly over three-quarters of the production throughout almost all of the 1730s. During the years of declining production after 1739 the resident fishery lost some of its lead due to its greater rate of decline and to a slight initial resurgence in the transient fishery. This pattern reversed itself in 1744 as the outbreak of war had a more detrimental effect on French fishermen than it did on colonial fishermen. However, missing data for the years 1741 and 1742 make these findings somewhat inconclusive. The single data breakdown for the second French period in 1753, however, reiterates the preeminence of the fishing proprietors.

The transfer of both resident and transient fishing efforts from Plaisance and the Chapeau Rouge coast to Isle Royale enabled the colony's fishery to attain high levels of production quickly. As shown in Table 10 the Isle Royale fishery reached, within a few years of its founding, production levels comparable to the highest it would ever attain. This did not mean that production in the fishery remained relatively static throughout the colony's history. The migrant fishery underwent a rapid decline during the early 1720s, and allowing for gaps in the information, it fluctuated at approximately one-half its former levels after the mid-1720s. The resident fishery gradually increased its own production to offset this decline but, as in all fishing endeavours, production was subject to annual fluctuations.

The supremacy of the resident fishery at Isle Royale was not wholly unexpected. After all, the migrant fishermen only prosecuted the summer fishery and indeed were forbidden by law from engaging in the winter shallop fishery.[2] It is not surprising that the fishing proprietors with their right to fish during both seasons were more productive than their more restricted French counterparts. During the 3 years from 1733 to 1735 the *commissaire-ordonnateur* credited the resident winter shallop fishery with an annual average of 19.6% of Isle Royale's total dried fish production.[3] Of course these production figures may have been exaggerated as they were obtained by merely multiplying the number of summer shallops by a per unit production figure. Statistics provided by the Admiralty, whose greater detail inspires more confidence, show that for the same triennia an annual average of only 68.0% of the resident summer shallops were employed in the winter fishery.[4] However, the Admiralty statistics do not contain per unit production figures, so that the *commissaire-ordonnateur* may have compensated for the lower number of shallops by simply using a lower multiplier.

The very denial of the winter shallop fishery to transients probably accounted for the marked difference in the capital fishing equipment used by residents and transients. As noted in the chapter on methodology there were two types of capital fishing equipment used at Isle Royale — schooners and shallops. Clark and Innis have noted marked differences in the ownership of these vessels, however.[5] As shown in Table 11[6] residents completely dominated the shallop fishery while transient fishermen were preeminent on board the schooners. More importantly, the relative

Table 10. Annual production of the resident and French migrant fisheries at Isle Royale, 1715–55, in *quintaux* and percentages

Year	Resident		French migrant		Total	
	qtx.	%	qtx.	%	qtx.	%
1715						
1716						
1717						
1718	76 000	48.6	80 500	51.4	156 500	100
1719	85 120	54.4	71 400	45.6	156 520	100
1720						
1721	78 000	62.1	47 600	37.9	125 600	100
1722						
1723	72 400	59.8	48 760	40.2	121 160	100
1724						
1725						
1726	106 100	75.3	34 800	24.7	140 900	100
1727	94 880	70.4	39 800	29.6	114 680	100
1728						
1729						
1730	125 060	75.5	40 570	24.5	165 630	100
1731	129 180	77.1	38 360	22.9	167 540	100
1732						
1733	123 470	74.7	41 895	25.3	165 365	100
1734	107 770	77.1	32 040	22.9	139 810	100
1735	109 205	76.6	33 290	23.4	142 495	100
1736	113 100	74.8	38 010	25.2	151 110	100
1737	120 950	81.0	28 350	19.0	149 350	100
1738	116 870	76.7	35 600	23.3	152 470	100
1739	100 300	69.8	43 360	30.2	143 660	100
1740	82 250	66.8	40 900	33.2	123 150	100
1741						
1742					83 410	
1743	61 920	69.8	26 800	30.2	88 720	100
1744	55 630	80.1	13 800	19.9	69 430	100
1750						
1751					93 580	
1752					83 130	
1753	78 940	80.2	19 510	19.8	98 450	100
1754						
1755						

parity in vessel ownership seen in the early years changed quickly to specialization by both residents and transients, with some annual fluctuations. The sole exception to this generalized rule occurred in the second French occupation when the decrease in the transient fishermen gave residents an advantage in the ownership of both schooners and shallops.

The explanation put forward for this variation in vessel type ownership was the notably higher capital cost of schooners.[7] Transient fishermen, with the backing of wealthy metropolitan outfitters, were better able to afford the expensive new technology. The more financially unstable fishing proprietor had to make do with the considerably less expensive

Table 11. Resident and transient ownership of shallops and schooners in the Isle Royale fishery, 1715-55, in percentages

Year	Shallops			Schooners		
	Residents	Transients	Total	Residents	Transients	Total
1715						
1716						
1717						
1718	48.6	51.4	100.0			
1719	54.4	45.6	100.0			
1720						
1721	55.8	44.2	100.0	50.0	50.0	100.0
1722						
1723	59.7	40.3	100.0	36.6	63.4	100.0
1724						
1725						
1726	78.1	21.9	100.0	33.8	66.2	100.0
1727	72.7	27.3	100.0	29.8	70.1	99.9
1728						
1729	86.3	13.7	100.0	34.7	65.3	100.0
1730	81.0	19.0	100.0	35.1	64.9	100.0
1731	82.4	17.6	100.0	28.4	71.6	100.0
1732	84.1	15.9	100.0	30.4	69.6	100.0
1733	86.4	13.5	99.9	35.8	64.2	100.0
1734	92.6	7.4	100.0	36.2	63.7	99.9
1735	88.8	11.1	99.9	34.2	65.7	99.9
1736	87.1	12.9	100.0	24.2	75.8	100.0
1737	95.6	4.4	100.0	22.6	77.4	100.0
1738	91.1	8.9	100.0	27.9	72.1	100.0
1739	87.2	12.8	100.0	33.3	66.7	100.0
1740	89.6	10.4	100.0	21.7	78.3	100.0
1741						
1742						
1743	91.8	8.2	100.0	26.9	73.1	100.0
1744	77.0	23.0	100.0	17.6	82.3	99.9
1750						
1751	63.1	36.9	100.0	64.8	35.2	100.0
1752						
1753	83.6	16.4	100.0	72.0	28.0	100.0
1754				93.7	6.2	99.9
1755						

shallop. Unfortunately this argument did not correlate to the demonstrated ability of the colonists to purchase schooners for the coasting trade. This also undercut arguments based on differences in operating costs. Although the schooner fishery necessitated a higher consumption of salt, it was uncertain how labour costs affected the higher producti-vity vis-a-vis the comparable unit of two shallops.[8] Prevost suggested that the occasional loss of an anchor and hawser made the schooner fishery more expensive.[9] This did not explain why transient fishermen would abandon a known technology for one with higher capital costs and possibly higher operating costs.

Unless there was some off-season use of schooners, in which transients had a marked advantage over residents, the answer to this paradox must have lain in the denial of the winter shallop fishery to transients. Per unit production costs had to be lower in the schooner fishery to justify the higher capital costs. Transient fishermen, unable to use their shallops in the winter fishery, accepted the higher capital costs of schooners in return for lower per unit production costs. Resident fishermen, on the other hand, presumably traded off slightly higher per unit production costs in the summer fishery for lower capital cost shallops. As these shallops were also used in the winter fishery and schooners were not, no matter who owned them, residents maximized their position by accepting the higher per unit production costs in return for the ability to participate in both summer and winter fisheries. Simply put, if a fisherman engaged only in the summer fishery a schooner was his best investment; if not, then it was the shallop. With a relatively consistent level in the mixing of ownership patterns, the financial advantages encouraging schooners cannot have been overwhelming. Of course, this reasoning remains speculative in the absence of detailed account books for both the schooner and shallop fisheries.

The favoured legal status of the resident fishery was not limited to participation in the winter fishery. In the distribution of land — a perennial problem in the dry fishery — residents had a marked advantage. Fishing proprietors received exclusive right to the shoreline concessions granted them — a sharp contrast with Ile Percee where the migrant fishermen had first choice.[10] This still necessitated regulating the distribution of unconceded shore property to the arriving seasonal fishermen. Initially all such fishermen were required to register their temporary establishments at Louisbourg, as was done at Havre du Croc on the Petit Nord. Another attempt resurrected the division of the coast into resident and transient areas as had been done at Plaisance and the Chapeau Rouge coast.[11] In this instance, Louisbourg harbour was reserved for residents while migrant fishermen were directed to Scatary and Menadou. Finally, the practice of designating the first fishing captain arriving at an individual harbour as the fishing admiral was adopted. The fishing proprietors of the colony were obviously exempt from the dictates of these admirals. This practice persisted, however, even in heavily settled fishing harbours. In 1743 Joannis Dalfoust applied to the *Amirauté* at Louisbourg for the privileges of fishing admiral at Niganiche, as he was the first arrival of the season there.[12]

Although the French migrant fishery could utilize or rent shorefront property anywhere along Isle Royale's coastline, these fishermen adopted relatively specialized areas of exploitation as time progressed. As Table 12[13] illustrates, initially there was a fairly even distribution of the migrant fishing effort all along the colony's coast with the inevitable concentration at Louisbourg/Scatary. Thereafter this latter area, particularly Louisbourg, increasingly became the focal point for the migrant fishery. Only Niganiche and, to an even lesser extent, Petit Degrat and Isle St. Jean proved sufficiently attractive to lure some vessels from the capital's embrace.

A number of reasons contributed to this concentration of the migrant fishery at the colony's larger fishing harbours. Perhaps the most important of these was the practice migrant fishermen had of combining both trading and fishing.[14] Recourse to Isle Royale's larger centres, particularly Louisbourg, ensured the cargo's sale. In addition, as resident fishermen appropriated the best shorefront properties migrant fishermen could no longer be certain that either unconceded or rentable property would be available in the smaller outports. Finally, as shallops and schooners were utilized for the actual fishing, it was necessary to secure a safe anchorage for their transatlantic vessels during the summer fishing season. The exposed harbour at Menadou, which was initially reserved for migrant fishermen, was abandoned by the early 1720s.[15] Vessels attracted to Niganiche by the excellent fishery and trading prospects were required to shift their anchorage to Port Dauphin, after 15 August, due to lack of shelter from autumn gales.[16] Louisbourg's large sheltered harbour, on the other hand, provided a safe refuge for a large number of vessels throughout the fishing season.

Although Isle Royale's resident population grew, the colony's human resources were never sufficient to meet the manpower requirements of the resident fishery. This fishery remained dependent on the seasonal supply of fishermen from France. Although some of these fishermen wintered in the colony, many returned to France after the summer fishing season. The La Roque census of 1752, for example, recorded the presence of some 200 fishermen

Table 12. Annual distribution of the migrant fishing vessels in Isle Royale for selected years 1718-53

Harbour	1718	1723	1727	1731	1735	1739	1753
Daspé		1	3				
Niganiche			4	4	4	5	3
L'Indienne	1	1					
Scatary	6	2	2	3			2
La Baleine	4	2	1	2	1		
Petit Lorembec				1			1
Louisbourg	12	38	23	35	16	26	10
Fourché							
Saint Esprit							
Isle Michaux							
Petit Degrat			3	3	1	2	
Isle St. Jean				1		2	
Isle Magdelaine							
Elsewhere	20[a]	6[b]	1[c]				
Total	43	50	37	49	22	35	16

[a] 6 at Port Dauphin, 4 at Menadou, 6 at Nerichat, 2 at Canseau, and 2 at Gabarus.
[b] 6 at Port Dauphin.
[c] 1 at Port Dauphin.

wintering in the colony's outports, exclusive of Louisbourg.[17] However, this figure represented well under half of the summer fishery labour requirements of the listed fishing proprietors. With such a dependence on the seasonal appearance of French fishermen, a shortage of labour loomed as a persistent spectre before the fishing proprietors. Indeed, in 1715 labour was in such short supply that many residents were faced with the prospect of economic ruin.[18]

Although severe shortages such as this appear to have been relatively rare, the potential for them encouraged abuses on both sides of the labour contract. Fishermen, aware of their relatively strong bargaining position, attempted to secure the highest wages and as many fringe benefits as possible. This included a basic wage paid in fish according to their function, a possible share in the oil and roe produced, and fringe benefits such as payment for passages to and from the colony or half passages if the fishermen were wintering in the colony. Some fishermen were known to sign with one fishing proprietor and then switch employers if additional bonuses were secured. For their part, fishing proprietors were not above befuddling a fisherman's head with drink prior to engaging him or offering the necessary inducements to lure a fisherman away from another employer.[19]

Beginning in 1718 and continuing until 1720, an attempt was made to regularize labour relations by requiring fishing proprietors to keep two registers — one for contracts and one for accounts.[20] Without these registers all contracts and accounts were to be considered void. Fishing proprietors were not allowed to engage a fisherman without him first presenting a congé or release from his previous employer. These releases were to be given to the fishermen immediately after the contract was completed and could only be withheld for debt if the debts were accrued in support of his family. In these instances, the fishermen were required to work off the debt but otherwise they were not. In this way fishing proprietors were prevented from keeping their fishermen by means of perpetual debt — especially for drink. To be considered binding all contracts had to be registered within 24 hours of signature. Due to the shortage of notaries in the outports this function could also be performed by the local curé or principal inhabitant. The registration of contracts was frequently ignored as much through lack of supervision as through the prevalent illiteracy. These provisions were reenacted in the comprehensive fisheries act in 1743.[21]

Other aspects of the relationship between the fishermen and the fishing proprietors were also regulated by law. As the fishermen and shoreworkers were paid in dried fish, it was the custom at Plaisance and later in Isle Royale for the fishing proprietor to have first opportunity to purchase these fish at the current price in the colony. This provided the fishing proprietors with a source of potential profit in the purchase and resale of these fish. In 1723 this custom received legal sanction and the fishermen and shoreworkers had to receive written permission from the fishing proprietors to sell to any other buyer.[22] The fishing proprietors enjoyed a much more complete legal advantage with regard to the selling of liquor to their fishing crews. While only the fishing proprietor could legally sell liquor to his crews, the frequent reiteration of this right in local ordinances indicates that it was far from scrupulously observed.[23] By the ordinance of 1743 fishing proprietors were only to advance two-thirds of their employee's anticipated salaries to them in such purchases.[24] Fishermen, on the other hand, were responsible for compensating the fishing proprietor for good fishing days lost because of drunkenness. This compensation was to be made on a pro rata basis from the catch of fishermen who had fished those days.

The only group portrait we have of the men who formed the transient pool of labour in the fishery is found in the La Roque census of 1752.[25] In February and March of that year Sieur De La Roque travelled the island of Isle Royale with the notable omissions of Louisbourg and Niganiche, enumerating the inhabitants and their dependents. In addition to identifying the individual fishing proprietors, La Roque also identified in most cases the "trente six mois hommes" or indentured workers, the servants and domestics, and the fishermen who were wintering with them. As the commitment of both the indentured workers and the servants and domestics to the fishery is somewhat questionable they have been excluded from the survey. These groups amounted to some 50 individuals in the employ of identifiable fishing proprietors. Although La Roque made reference to approximately 250 fishermen who were wintering in the colony, names, ages, and place of origin are given for only some 200 of them. The age distribution of this group appears in Table 13[26] and the places of origin are given in Table 14.[27] A rough estimate of the summer

Table 13. Ages of *compagnon* and *garçon pêcheurs* wintering in Isle Royale outports, February/March 1752 (200 cases)

| Age group | *Compagnon* and *garçon pêcheurs* | | |
	Real No.'s	Percentages	Cumulative percentages
10-14	2	1.0	1.0
15-19	18	9.0	10.0
20-24	53	26.5	36.5
25-29	34	17.0	53.5
30-34	31	15.5	69.0
35-39	18	9.0	78.0
40-44	16	8.0	86.0
45-49	11	5.5	91.5
50-54	10	5.0	96.5
55-59	5	2.5	99.0
60-64	1	0.5	99.5
Over 65	1	0.5	100.0

Table 14. *Province* and *evêché* of origin of *compagnon* and *garçon pêcheurs* wintering in Isle Royale outports, February/March 1752 (199 cases)

Province and evêché of origin	Province of origin		Evêché of origin	
	No. of cases	%	No. of cases	%
Normandie	44	22.1		
Rouen			2	1.0
Coutances			18	9.0
Avranches			22	11.0
Unidentified			2	1.0
Bretagne	40	20.1		
Dol			2	1.0
Rennes			2	1.0
St. Malo			26	13.1
St. Brieux			4	2.0
Tréquier			1	0.5
Nantes			3	1.5
Unidentified			2	1.0
Aunis	3	1.5		
La Rochelle			3	1.5
Gascogne	97	48.7		
Bayonne			19	9.5
St. Jean de Luz			78	39.2
Ile de France	1	0.5		
Paris			1	0.5
Marche Limousin	2	1.0		
Lemoges			2	1.0
France (unknown)	4	2.0	4	2.0
Colonies	8	4.0		
Isle Royale			6	3.0
Plaisance			1	0.5
Canada			1	0.5

work force required for the fishing vessels identified in this census would fall in the range of 500-700 men.

As might be expected, given the arduous nature of the work, most fishermen were in the prime of life. Almost 70% of the 200 fishermen whose ages were identified by La Roque were under the age of 35. If the age limit is increased to include those under 40 years of age this group would include fully 78% of all the fishermen. At the same time only 1% were under the age of 15 and only 10% were under the age of 20. Of course, the timing of the census may have minimized the numbers of these younger age categories. During the fishing season the shore work offered employment opportunities for young boys in such chores as turning the fish. It is

uncertain, however, how many of these jobs may have been filled by the younger family members of the fishing proprietors.

Similarly, the place of origin for most of these fishermen offers few surprises. As might be anticipated the coastal provinces of France provided the majority of fishermen. In particular the three provinces of Normandy, Brittany, and Gascony accounted for 90.9% of the 199 fishermen whose origins were identified.[28] Even within those provinces the actual places of origin were quite clustered. In the north the seven dioceses of Coutances, Avranches, Dol, Rennes, St. Malo, St. Brieux, and Tréquier, all relatively near St. Malo, produced 37.6% of these fishermen. In the south the two largely Basque dioceses of Bayonne and St. Jean de Luz accounted for another 48.7%. Surprisingly, colonial-born fishermen constituted only a very small proportion of the total. Isle Royale accounted for only 3% while Plaisance and Canada accounted for only half a percentage point each. In part, the shortness of the colony's existence explains this, but perhaps even more important would have been the greater opportunities of the colonial-born for occupations outside the fishery.

Throughout Isle Royale's existence the French also practised the dry fishery at other locations and these areas competed with Isle Royale for the fishermen's attention. For over 20 years before the first siege Cape Ray in Newfoundland acted as an alternate, if officially discouraged, base for the colony's fishermen. In 1723 the Récollet missionary at Scatary warned the officials at Louisbourg that a vessel captain from Granville planned an expedition to Cape Ray.[29] Rousseau de Souvigny, an officer of the garrison detached to Scatary, published an order that September prohibiting fishing proprietors from going to Cape Ray.[30] Several were stopped but one sent two vessels with 30 men, which would be brought back in the spring. The Minister of Marine complimented the officials on their actions but the problem was not corrected as easily as that.[31] In September 1724 the *commissaire-ordonnateur* learned that the Granville captain and several others had gone to Cape Ray; one of them had received permission from the governor on condition he return but had since died.[32]

Cape Ray continued to attract French fishermen during the 1730s but also acquired a bad reputation. In 1730 an Englishman named Richard established himself at Isle des Graules near Cape Ray where a number of men from Ingonish joined him. To complicate matters Richard imprisoned 15 Micmacs from Isle Royale which caused their comrades to want to attack the English in revenge. Louisbourg officials feared Richard's followers would become pirates and pillagers and complained to the English commander at Canso.[33] A year later the French still had not received a reply. The Cape Ray group, on the other hand, were reported increasing in numbers daily and were enticing fishing crews from Scatary to desert their masters.[34] By 1733 French merchant vessels going to Isle Royale were reportedly stopping to trade with the "brigands" at Cape Ray.[35] This practice was forbidden but its scale was small as that year's total was only one vessel from St. Malo and three Basque vessels which ordinarily fished there.[36] The prohibition on French vessels trading there was not completely adhered to as in 1737 two sailors from Louisbourg hoped to go to Cape Ray to secure a passage on a vessel returning to France.[37]

The reasons for Cape Ray's continued attractiveness were a blend of economic opportunity and freedom from restraint. In 1733 Maurepas stated that many people went to Cape Ray to avoid certain unidentified costs in the purchase of fishing commodities at Isle Royale.[38] Louisbourg officials replied it was the commerce of French vessels on this coast of Newfoundland which occasioned fishing crews to desert the colony.[39] In addition many individuals sought refuge there from French law. In 1730 fishing crews from Ingonish had reportedly stolen from their masters and then deserted to Isle des Graules.[40] Jacques Massé helped steal a fishing shallop at Louisbourg in 1737 with the plan of sailing to Cape Ray to secure a passage back to France on a vessel leaving there.[41] Unfortunately, no estimates were provided for the number of French who eventually established at Cape Ray.

Commissaire-ordonnateur Bigot visited Ingonish in 1742 and discovered a great liaison between there and Cape Ray. Frenchmen from there came to Ingonish to get married and then returned to Newfoundland with their brides. Indeed, he had seen two newly married Ingonish women who had come from Cape Ray who planned to rejoin their husbands there. Bigot ordered his subdelegate not to let them depart; they had to get their husbands to return to fish in the colony. Furthermore he asked the priest not to perform any more of these marriages. Bigot noted that the number of French at Cape Ray was increasing as the fishery was abundant and they were independent there.[42] Although an undetermined number of individuals left Isle Royale for Cape Ray, most participants in the colony's fisheries appear to have been successful and contented in their occupation.

FINANCES AND THE FISHERY

Central to the prosecution of the fishery at Isle Royale was the role of the *habitant-pêcheur* or fishing proprietor. This individual acquired by purchase or rental a fishing property where he gathered the buildings, equipment, supplies, and specialized labourers needed for efficient production.[1] Naturally these necessary units of production could only be assembled through the expenditure of capital and hence the ability to raise capital was a prerequisite for the budding fishing proprietor. Generally speaking, fishery costs were divided between relatively fixed investment in land and capital equipment and seasonal operating costs for supplies and labour. The existence of rental markets for both land and capital equipment helped minimize investment costs, but this saving was offset by the higher operating costs which renting created.

As noted in preceding chapters, improved fishing properties and items of capital equipment such as schooners or shallops necessarily entailed large expenditures on the part of the fishing proprietor. Fishing properties varied in value from one to several thousand *livres* according to size, location, and the degree of development. The value of schooners fell in this same price range, again varying according to size and condition. Shallops were markedly cheaper such as those Louis Gilbert contracted to build in 1751 at 350 *livres* each.[2] Although the shallop fishery required less capital expenditure and produced a higher quality dried fish than the schooners, production per man was lower. In spite of this the lower capital cost undoubtedly encouraged new fishing proprietors to begin their career in the shallop fishery.

There were additional methods, however, of minimizing the initial capital expenditure needed to enter the fishery. Perhaps the most obvious means for the new fishing proprietor was to start on a small scale and expand his involvement as opportunity permitted. In this manner the initial purchase need only be an undeveloped fishing property which could then be gradually improved with the help of some hired labour. The rental or purchase of used shallops and schooners also lessened the expense of these items. For example, throughout most of Louisbourg's existence the availability of older New England schooners provided a cheaper alternative to the purchase of new vessels. In his 1752 census of Isle Royale, La Roque noted a Laurembec resident who used both methods of reducing the cost of entering the fishery.[3] Jacque Couzin, aged 26 and married with three small children, had only one indentured servant to help him clear his fishing property. Couzin had been verbally granted his concession by Desherbiers and Prevost and had cleared one *arpent* and had made a stage and flakes "pour la secheru de deux chaloupes." Although he did not own a shallop Couzin hoped to rent one for the coming season. Clearly, here was a man attempting to become a fishing proprietor with a minimum of financial backing.

The fishing property developed by Couzin represented a minor investment compared with that of a larger, more established operation. In July 1733 Joannis Daccarrette and Eustache Lagarande Lepestour appraised the buildings alone on Bonnaventure LeBrun's fishing property on Ingonish Island at 3130 and 3360 *livres*, respectively. As shown in Table 15[4] the amount invested was not divided equally among the buildings. In particular there was great discrepancy between the value placed on the fishermen's accommodations and that placed on the house. With the exception of the forge, even the work and storage areas were more valuable than the fishermen's living quarters.

A 1754 inventory of a fishing property for the use of two shallops situated on the Magdalen Islands provides a more detailed account of the cost of establishing a small fishing property.[5] The buildings consisted of a cabin for the master with an adjoining shed for supplies, a storehouse for salt, a cabin for the fishermen, also with an adjoining shed, a fish stage, two wash cages, and a table for washing the cod. Three appraisers considered these to be worth 2200 *livres*, whereas a fourth valued them at 3000 *livres*. A list of expenses for the construction noted planks at a cost of 317 *livres* 1 *sol*, nails at a cost of 370 *livres* 10 *sols* 6 *deniers*, and iron at 3 *livres* 8 *sols*. Glass, locks, transportation, and wages were mentioned as other expenses but no costs were given. It would seem likely that purchased construction materials accounted for some 700-750 *livres* of the final appraised value of 2200 *livres*.

Three reviews of the summer shallop

Table 15. Appraisals of Bonnaventure LeBrun's fishing property on Ingonish Beach Island by Joannis Daccarette and Eustache Lagaranche Lepestour, 1733 (in *livres*)

Building	Daccarette	Lespestour
The fish stage	500	600
Two cabins for the fishermen	300	350
The cabin for the forge	160	180
The cabin with dry storage and a storehouse	900	800
The new house	1500	1200

Table 16. Review of operating costs for two shallops in the summer fishery for 1733, 1739, and ca. 1740 (in *quintaux* of dried cod)

Cost	1733	1739	ca. 1740
Labour	360.00	359.0	357
Shallop supplies	73.33	37.5	48
Fishing supplies	127.00	119.5	124
Food supplies	45.00	72.0	72
Property rentals	20.00	-	-
Total costs	625.33	588.0	602
Anticipated production	600.00	600.0	600
Profit (+) or Loss (-)	-25.33	+12.0	-2

fishery written between 1730 and 1740 compared operating costs with a projected total production.[6] In each instance a production unit of two shallops with a hypothetical return of 600 *quintaux* of dried cod was used. None of the assessments was designed to bolster investor confidence as the first detailed a cost overrun of 4.2%, the second a profit of only 2.0%, and the last a loss of 0.3%. As shown in Table 16[7] labour emerged as the greatest expense, accounting for 60.0, 59.8, and 59.5% of production, respectively, in the three reviews. Food, shallop, and fishing supplies combined to account for 40.9, 38.2, and 40.8%, respectively, of the projected returns. Only the first review allowed 3.3% of the anticipated production for the rental of beach flakes and cabins.

In part these seemingly bleak economic prospects in the fishery can be attributed to the motivations of the originators of these accounts. The first was written in 1733 by *ancien habitant-pêcheurs* to protest the abuses caused by intense competition for labour and to seek legislated wage controls, particularly for the fringe benefits offered fishermen. Naturally it was in these fishing proprietors' interests to paint as bleak an economic picture as possible. The second was prepared in 1739 by *commissaire-ordonnateur* Le Normant to show the need for legislated reforms to correct abuses in the fishery. At the time Le Normant was in France on his way to a new appointment in Saint Dominique (Haiti) so he seemingly had little reason to exaggerate the plight of the fishing proprietors.[8] Consequently he provided the most favourable assessment of the fishery but even he allowed only 2.0% profit over operating costs. The last review is contained in an unsigned memoir misdated as 1758. Internal evidence suggested it was written in 1739 or the early 1740s. Certainly it could not have been written after 1743 or the wage scale for fishermen would

Table 17. Wages on the shallop fishery "sur le pied de 600 quintaux pour deux chaloupes" at Plaisance and Isle Royale, various years, in *quintaux*

	Plaisance	Isle Royale				
	pre-1713	1715	1733	1739	ca. 1739/40	1743
Fishermen						
Shallop Master	36	40	38	38	38	36
Companion fisherman			36	36	36	32
Bosun	34	38				
Stower	32	36				
Shoreworkers						
Shoremaster	40	45	45	45	45	40
Salter			24	24	24	20
Header				20	36	17
Boy			18	18		13
Wages for 10 men						
for two shallops			325	327	325	290
Oil and roe	1/3	3/3	3/3	1/2		1/2
Passages			35	32	32	

have followed the regulation of that year. Once again, the purpose of this review was to illustrate abuses in the fishery.

As illustrated in Table 17, labour constituted a major expense to the fishing proprietor in the prosecution of the fishery. In 1733 the *ancien habitant-pêcheurs* complained that wages accounted for 54.2% of total dried fish production in the summer shallop fishery. Le Normant credited labour costs as 54.5% of production in a description of the same fishery in 1739. The anonymous account written at approximately the same time allowed 54.2% of production as wages. When the allowances made for passages were included, these proportions increased to 60, 59.8, and 59.5%, respectively. This uniformity in labour costs is explained by the fact that labour was paid *pro rata* to production. Although it was common to speak of the pay for various positions as being so many *quintaux,* this was done "sur le pied de 600 quintaux pour deux chaloupes." The worker's wage was actually a proportion of total production. For example, Noel Lechanche was hired for the summer fishery in 1734 "sur le pied de 18 qx. de Molue sur 600" and received 12.95 *quintaux* as his share of a total production of 431.67 *quintaux*.[9] Provision was also made for the division between the fishing proprietor and his crew of the fish oil and roe produced. The development of the winter shallop and the schooner fisheries led to modifications in the allotment of wages.

As the basis for the wage systems in the other fisheries, the pay allotments in the summer shallop fishery merit particular attention. The system had its origins in Plaisance

and the usage of 600 *quintaux* for two shallops probably reflected the average catch there.[10] Indeed, English records of the time reported the French to be landing 300 *quintaux* per four-man shallop.[11] This practice was retained at Isle Royale as a convenient traditional means of accounting. As shown in Table 17,[12] the share varied according to the responsibility, skill, and labour involved in the work being done.

As their wages were proportioned to the catch fishermen sought to obtain as large a share as possible. In 1715 Saint Ovide reported share increases for the shoremaster, shallop master, bosun, and stower over the former rates at Plaisance due to a labour shortage in Isle Royale.[13] These increases raised labour costs for two shallops by 29 *quintaux* or 4.8% of the projected production. Later assessments of the fishery show that these labour costs were not a temporary phenomenon. Similar demands were noted with regard to fish oil and roe. The production of these items required extra involvement from the crew and it was customary to divide the proceeds between the fishing proprietor and crew. At Plaisance the workers reportedly received one-third of the oil but by 1715 they were demanding all of it. Later accounts, such as Le Normant's in 1731, indicated this share had been standardized at one-half.

In addition to increasing their share demands the fishermen tried to get other benefits included in their contracts. On 1 November 1733 Joannis L'Etapy signed an engagement with François Lessenne to act as salter at Lessenne's fishing property at St. Esprit for both the autumn and summer fisheries. His engagement specified a weekly allowance of a *pinte* (0.93 L) of rum in addition to his regular share of 22 *quintaux* of every 600 produced.[14]   In 1733 the established fishing proprietors complained that competition for labour from new fishing proprietors necessitated the payment of passages for fishermen coming from France, as well as for those wintering in the colony. These passages were paid at the rate of 3.5 *quintaux* for each of the 10 workers needed for two shallops or some 5.8% of total production. Later accounts indicated full passages (at a cost of 4 *quintaux)* were reserved for fishermen coming from France while those wintering in the colony received half passage. The header and the boy were dropped from the roll of those receiving these benefits. Passage money varied from 32 *quintaux,* if all the fishermen came from France, to only 16 *quintaux* if they all wintered in Isle Royale.

The generally small scale of operations of individual fishing proprietors created a competitive labour market which rendered group action to limit wage rates and fringe benefits unlikely. Instead, fishing proprietors depended on government legislation to control the wage rate. The first such regulation was designed to restrict the competition for the colony's scarce supply of labour. In 1721 the governor and *commissaire-ordonnateur* passed an ordinance prohibiting foreign merchants from engaging the fishermen of the inhabitants.[15]   Although this, in theory at least, protected the fishing proprietor from outside encroachments on the labour supply, the problem of competition from other colonials was left unresolved. Finally, in 1733, the long-established fishing proprietors complained of the excessive fringe benefits, principally paid passages, being offered by newcomers. In September of that year an ordinance, which was repeated the next year, forbade the payment of half or full passages or any other benefit beyond the "ancien usage de cette colonie."[16]

The culmination of this attempt to restrict legally the wage levels of fishermen and shore workers came in 1743 with the issue of a comprehensive fisheries ordinance for Isle Royale. Within the summer shallop fishery, wages were restricted to 290 *quintaux* or 48.3% of total production at the rate of 600 *quintaux* for two shallops. Individual wages varied from 40 *quintaux* for the shoremaster to 13 *quintaux* for the boy. The crew's share of the oil and roe was similarly restricted to one-half of the total, with this half being divided proportionally among the crew members according to their wage rate.[17] Thereafter, both the fishing proprietors and their labour force had to abide by the provisions of this act in their engagements. Bonuses could neither be given nor received. Indeed, it was common practice during the second French occupation to note in the contract that the salaries were those prescribed by the 1743 act. Although most engagements after 1743 specified the wages stipulated in the fisheries ordinance of that year, exceptions did occur. Each of three fishermen hired by Jean Milly in July 1749 were engaged under different terms.[18] The first, hired as a fisherman from 19 July until the end of the next September, was to be paid at the rate of 30 *livres* per month. A second, hired as a companion fisherman for the same length of time, was to be paid according to the regulation — presumably that of 1743. The third, hired 9 days later but to end employment at the same time, was to receive half the fish he caught. In this instance, Milly was only responsible for providing the lines. As indicated by the above examples, the end of September marked the end of the summer fishery and the time for payment of both labour and supplies. Indeed, 29 September, St. Michael's Day, became the customary day in Isle Royale for weighing the summer's production and the settlement of debts, including labour.[19]

Labour contracts were made to cover a variety of time periods. In the shallop fishery there was a summer season lasting from May until the end of September and a winter season from mid-November until February. Some fishing proprietors hired their fishermen for one season; others hired for both. In the winter fishery, fishermen were hired at the same rate as the summer fishery or on an equal division of the catch between the fishermen and the fishing proprietor.[20]   In the schooner fishery fishermen received four-sevenths of the catch, to be divided among themselves, and the fishing proprietor received the remaining three-sevenths. The different seasons and types of vessels led to occasional mixing of terms and remunerations in the contracts. In 1733 Joannis L'Etapy agreed to serve François Lessenne as a salter in the winter shallop fishery on the equal division basis and in the same capacity during the summer on board a fishing schooner for a specified share.[21]

In addition to the fisherman's share of the catch, the fishing proprietor also provided him with room and board. The foods given were the most basic staples — bread or biscuit, dried peas, butter, salt pork, molasses, and rum or brandy. As illustrated in Table 18,[22] food costs for the crews of two shallops in the summer fishery were estimated at between 45 and 72 *quintaux* of dried cod. Unlike labour, food costs were relatively fixed and would not have varied according to production. In terms of quantity the foods given the fishermen appear to have been in amounts similar to those given soldiers and sailors.[23] During times of food shortage in the colony the supplies issued the fishermen could be rationed. On 6 May 1742 Duquesnel and Bigot passed an ordinance restricting the daily ration given each "compagnon, matelot, et Engage" to 1 *livre* of bread and the normal allotments of salt pork, butter, and brandy.[24] The arrival of supply vessels from Quebec and New England later that summer permitted the resumption of the full bread and vegetable ration.[25]

The purchase in the fall of 1738 of biscuit, rum, molasses, and peas by François Gautier and his companions for their winter supplies confirmed the lack of variety in the basic staples.[26] The monotony of such a diet encouraged fishermen to supplement it whenever possible. Fresh fish were an obvious and readily available foodstuff and were undoubtedly utilized by the colony's fishermen. In 1738 it was reported that the soldiers, who were issued essentially the same type of rations, ate mainly fish.[27] The gardens commonly found on fishing properties provided another source of food supplies. René Herpin, for example, had two gardens on his small fishing property on Louisbourg's north shore.[28] *Commissaire-ordonnateur* Prévost reported in 1753 that fishermen coming from France were supplied with biscuit, beans, oil, brandy, and cider on the voyage over. Fishermen supplemented this diet with purchases of eggs, cheese, sardines, hams, and brandy.[29] Isle Royale fishermen followed a similar practice of purchasing foodstuffs from fishing proprietors and others. The estate of Marlan Duhart, who drowned at Fourchu in 1734, included his personal supplies of Holland cheese, ham, and Basque bread.[30]

Particularly in the smaller establishments, cooking arrangements were probably made informally among the shore workers and fishermen themselves. The presence of bake

Table 18. Cost of food supplies for two shallops in the summer fishery as reported in reviews for 1733, 1739, and ca. 1739, in *quintaux*

Food	Cost		
	1733	1739	ca. 1739
Biscuit or bread	25	40	25
Peas	6	8	5
Butter	6	8	16
Salt pork	-	5	6
Molasses	4	3	4
Rum or brandy	4	8	16
Total	45	72	72

ovens on some of the larger properties suggests that an increased scale of operations may have led to more formalized cooking arrangements. As in the military, stews probably were one of the simplest and most common meals prepared by the fishermen. The only cooking utensil François Blondel provided to the fishermen staying in his cabins was a copper *chaudière*, or cooking pot, of some 56 L capacity for making soup.[31] Of course, this *chaudière* or even a larger one could have been used for brewing spruce beer. The molasses ration would normally have been brewed into this beverage. Especially in winter, the anti-scorbutic properties of this brew compensated for the vitamin deficiencies of the dried and salted foods. Hard liquor was probably issued in a weekly allowance as suggested by some contracts.

Like food supplies, fishing supplies formed a relatively fixed cost to the fishing proprietor. Regardless of the level of production, the fishing proprietor had to purchase lead, lines, and hooks so his men could fish, mackerel and herring nets and a shallop for tending them to provide bait, nails to repair his fish stage, and salt to preserve the catch. As shown in Table 19,[32] these costs were estimated in the 1730s as totalling between 119.5 and 127 *quintaux* of dried fish. Salt comprised the greatest expense with either 70 or 80 *barriques* of salt allotted for a production of 600 *quintaux*. The price of a *barrique* of salt was consistently quoted at 1 *quintal* of dried fish. However, if production was low, some salt would be left

Table 19. Cost of fishing supplies for two shallops in the summer fishery as reported in reviews for 1733, 1739, and ca. 1739, in *quintaux*

Supplies	Cost in *quintaux*		
	1733	1739	ca. 1739
Nails for stage	-	-	2
Lines	8	9.0	10
Lead	3	4.0	4
Hooks	3	2.0	3
Herring nets	21	32	14
Mackerel nets	12		11
Shallop for bait	-	2.5	-
Salt	80	70.0	80
Total	127	119.5	124

over for the next fishing season. Many of the other items, such as lines, leads, and nets, would be used or lost through the act of fishing, whether or not fish were actually caught. The shallop crews also had to be provided with baskets and buckets to hold their daily provisions and compasses to ensure safe navigation in foggy weather.

Shallop supplies were another source of relatively fixed costs for the fishing proprietor. The beginning of each fishing season required that the shallops be put in a state of good repair. The shallops themselves were considered to last only four seasons and the fishing proprietor had to include their depreciation in his costs. Both old and new shallops were to be found on Isle Royale fishing properties.[33] There were also annual costs for tar, caulking, and nails to repair the hull, as well as the replacement of broken operating gear. This latter group included the sails, rigging, grapnels, hawsers for the grapnels, and a cable to moor the shallop. A less familiar expense were the *falques* or collapsible canvas washboards attached to the shallop's gunwale. In heavy seas or when fully laden, these washboards were raised to prevent the shallop from taking water. As shown in Table 20,[34] these costs were estimated at between 37.5 and 73.3 *quintaux* per year. To ensure the crew took proper care of their equipment the 1743 ordinance made the fishing crews responsible for a third of any

damage to their shallops, rigging, or gear.[35]

It seems likely that the three fishing reviews exaggerated the costs of food, fishing, and shallop supplies to underscore the occasionally financially precarious position of the fishing proprietor. For example, the fishing proprietors allotted in their 1733 memoir the total cost of the fishing shallops rather than quartering it to the actual depreciation.[36] In an age before synthetics, natural fibre ropes, nets, lines, and sails wore out considerably faster than their modern counterparts. However, all three memoirs appear to have allowed an extremely high replacement rate for fishing gear. Unfortunately, the absence of detailed business records prohibits a finer assessment of the accuracy of the replacement rate. Christopher Moore determined that supplies would cost between one- and two-fifths of the annual earnings. In a small fishery such costs could easily exceed a third, whereas a larger fishery might reduce this to a fifth, presumably through economies of scale.[37]

A fishing proprietor's income came from a variety of sources related to his participation in the fishery. Perhaps his most obvious source of income was the fish remaining after he had paid the wages or shares of his fishermen and the debts for his necessary fishing supplies. In addition, the fishing proprietor

Table 20. Cost of shallop supplies for two shallops in the summer fishery as reported in reviews for 1733, 1739, and ca. 1739, in *quintaux*

Supplies	Cost in *quintaux*		
	1733	1739	ca. 1739
Shallops	40.00	7.5	10
Sail cloth	8.33	14.0	10
Grapnels	4.00	3.0	3
Hawsers	8.00	8.0	8
Cable	-	-	5
Washboards	3.00	2.0	3
Cordage	2.00	-	2
Tar, caulking, nails	4.00	-	4
Oars	4.00	3.0	3
Total	73.33	37.5	48

was legally entitled to one-half of the oil and roe produced in the shallop fishery and four-sevenths of this production in the schooner fishery.[38] Of course, the typical value of this residue was subject to different interpretations. In 1733 the fishing proprietors complained that in the summer shallop fishery, production costs outstripped the returns, a contention supported by a later anonymous memoir. Le Normant estimated in 1739 that the profit margin for this same operation was only 2.0% of the anticipated production.[39] It is probable that these figures were biased against the fishing proprietors as two were arguing for wage controls. The repeated concern over high wages, however, suggests that the profit margin in direct fishing operations was not always sufficient. This is particularly significant as many of the fishing proprietor's costs were relatively fixed for each fishing season, whereas production levels, and hence profits, fluctuated according to a variety of factors.

A less obvious source of income for the fishing proprietor was the profit derived from the purchase and resale of the wages of his fishermen and shoreworkers. These wages or shares were contracted in amounts of fish, oil, and roe and the fishing proprietor had first preference of purchase at the "prix courant de la coste" according to the custom at Plaisance and later at Isle Royale. On 8 September 1724 this custom was given legal basis by an ordinance passed by St. Ovide and De Mesy.[40] To sell to other buyers the fishermen had to have written permission from their masters. To prevent disputes over the "prix courant," 13 days after the first ordinance the price was set at 12 *livres* "en argent ou Bonnes Lettres, lad. morue marchande, et de refaction, mais non Compris celle de rebut."[41] Although French documentary evidence of further price setting has not yet been found, Warren claimed in 1739 that the French government annually set the price at around 9 shillings (approximately 10 *livres*).[42]

The fishing proprietors also had the sole right of providing drink to their fishermen. Although they provided some liquor to the fishermen as part of their ration, this undoubtedly followed the practice described by Nicholas Denys of being heavily watered.[43] Fishermen were noted as heavy drinkers and Le Normant reported that the times of the year between fishing seasons were times of debauchery.[44] Fishing proprietors took advantage of this predilection for strong drink by selling their crews large quantities of liquor. The fishing proprietors were not supposed to charge excessive prices but this was not always the case. In 1743 François Picard of Saint Esprit sold rum to his fishermen at almost twice the price it sold in Louisbourg.[45] The repeated *ordonnances* making such sales the exclusive privilege of the fishing proprietors indicates they had plenty of competition. About 1733 François Lessenne *dit* Francoeur requested the Superior Council to order M. Col to close his inn on an adjoining Fauxbourg property, as Lessenne's fishermen were spending all their time there. Lessenne argued that if the Superior Council was not prepared to do this, he should be given permission to stop his fishing operations and open his own tavern.[46]

The fishing proprietors also had the preference in providing clothes and other necessities, but again sales were to be made "au prix commun du détail dans la colonie."[47] This necessitated fishing proprietors keeping additional stock on hand for sale to their men. In the inventories of Elie Tesson La Floury were "huit Casaques etonse devantaux de curre," "quatre paires debottes neuves," and, perhaps more significantly, "vingt Cinq chapeaux cummum," all presumably for sale to his fishing crews.[48] François Picard's fishermen purchased "aunes d indesne, 5 paires de bas, un mouchoir chapau, cinq livres tobacq chemise, paire de botte."[49] Although Picard's fishermen did not make any purchases of food, such purchases would have taken place. In the inventories of fishing proprietors it is impossible to know which food items were for sale, which were to be given as rations, and which were for his personal use.

In some fishing operations the sale of liquor and other goods to the fishing crews was probably the dividing line between profit and loss. The death of the widow Perré awakened a deep concern among her heirs over the appointment of efficient managers to oversee the fishing business. As they noted in their brief to the court: "il Est de la Connoissance de Tout les habitants pescheurs qu'il nya que les proffit qu'ils peuvent faire sur les fournitures faites a leurs compagnons."[50] In this instance the administrators successfully concluded the fishery initiated by the widow Perré. However, in addition to the profit from actual fishing operations, her heirs also received some profit from over 1050 *livres* worth of goods sold to the crews of her eight shallops during the summer fishery. Although the actual profit margin has not yet been determined, the volume of sales indicates its utility as an income supplement to the fishing proprietor.

Fishing proprietors made use of all available sources of income in their annual efforts to end the fishing season with a profit. While some contemporary reviews of the fishing industry indicated that profits were minimal at best, these reviews neglected some sources of income and appear to have exaggerated some costs. The readiness with which individuals continued to enter the fishery throughout Isle Royale serves as an indicator that money could be made. Nevertheless, the fishery remained a precarious calling and a poor season often left fishing proprietors in debt and even bankrupt.

CONCLUSION

The cod fishery was the economic base of Isle Royale. Throughout its short history the supply needs of the fishery and the production of dried cod and fish oil dominated the colony's import and export trades. Although Isle Royale was but one of France's North American cod producing areas, the value of its production in most years easily exceeded that of Canada's fur trade. Indeed, dried fish exports formed a crucial link in the establishment of a triangular flow of goods between France, Isle Royale, and the French West Indies. Although Louisbourg's fortifications were the easternmost of a continent-wide ring hemming the English colonies to the Atlantic seaboard, their primary function remained the protection of Isle Royale and its valuable cod fishery.

The immediate spur for the French establishment of Isle Royale was the loss of Acadia and the south coast of Newfoundland to the English by the Treaty of Utrecht in 1713. The historic background to Isle Royale's settlement was the development of the dry fishery. Initially Europeans had exploited the North American fishing grounds through the green fishery, in which the catch was taken back to Europe in salt bulk without land even necessarily being sighted. The better preservation and handling qualities of dried cod gave rise to a dry fishery in which the catch was first dried on land before the return to the mother country. In spite of its disadvantages, the market for wet salted fish remained strong and throughout Isle Royale's history, the green fishery remained the most important branch of the French cod fishery.

The importance of land in the dry fishery led to the rapid development of the North American shoreline near the fishing grounds. Frequently this development led to bitter disputes over control of the best locations, a rivalry not confined to different nationalities but also between resident and migrant fishermen. At Isle Royale the French followed their earlier practice at Placentia and favoured resident over migrant fishermen. Initial grants of property were free and thereafter became the private property of the owner. A strip development was soon found on the shoreline of suitable harbours as each fishing proprietor sought to have his own water frontage. This strip development was also followed in the colony as a whole, with settlement primarily being located along the island's Atlantic coast.

Production within Isle Royale was restricted almost entirely to dried cod; by-products included cod liver oil and salted tongues and sounds. A small quantity of wet salted fish was also produced towards the end of the fishing season. Fishing was done from shallops going daily to the inshore fishing grounds and from schooners making longer voyages to the offshore banks. After cleaning and splitting, the catch was salted before starting a sometimes lengthy drying process. The dried cod was marketed in Europe and the French West Indies.

Central to the prosecution of this fishery was the role played by the fishing proprietor. This individual acquired by sale or rental a fishing property where he assembled the needed factors of production. Labour and supplies proved the greatest costs and in poor seasons the returns were not always able to cover expenses. Income came from a variety of sources, however, and the fishing proprietors proved moderately successful throughout Isle Royale's existence. Overall, the fishery remained a strong base on which to anchor the colony's fortunes.

APPENDIX A. REFERENCES FOR ISLE ROYALE FISHERY RETURNS

1716 - France. Archives Nationales, Archives des Colonies, Archives de la France d'Outre-Mer, Sér. G1, Vol. 446, Pièce 54. Récensement général de l'Isle Royale, 1716.

1718 - France. Archives Nationales, Archives des Colonies (hereafter cited as AN, Col.), Sér. C11B, Vol. 3, Fols. 206-8. St. Ovide au Ministre, 17 décembre 1718.

1719 - AN, Col., Sér. C11B, Vol. 5, Fols. 43v-44. St. Ovide au Conseil, 29 novembre 1719.

1721 - AN, Col., Sér. C11C, Vol. 15, Pièce 210. Conseil au St. Ovide et de Mesy, 7 décembre 1721.

1723 - AN, Col., Sér. C11B, Vol. 6, Fol. 245. Etat des navires qui sont venus, 12 décembre 1723.

1726 - AN, Col., Sér. C11B, Vol. 8, Fol. 230. Isle Royalle, 1726.

1727 - AN, Col., Sér. C11B, Vol. 9, Fol. 259. Isle Royalle, 1727.

1729 - AN, Col., Sér. C11B, Vol. 10, Fol. 211. Liste generale des Batteaux, Gouelettes et Chaloupes, 16 décembre 1729.

1730 - AN, Col., Sér. C11B, Vol. 11, Fol. 69. Isle Royalle, 4 décembre 1730.

1731 - AN, Col., Sér. C11B, Vol. 12, Fol. 64. Isle Royalle, 1731.

1732 - AN, Col., Sér. C11B, Vol. 13, Fol. 242. Liste generale des Batteaux, Gouelettes et Chaloupes, 22 décembre 1732.

1733 - AN, Col., Sér. C11B, Vol. 14, Fol. 232. Isle Royale 1733; and AN, Col., Sér. C11B, Vol. 14, Fol. 234. Estat des Batteaux, Gouelettes et Chaloupes, 1733.

1734 - AN, Col., Sér. C11B, Vol. 16, Fol. 257. Isle Royalle, 1734; and AN, Col., Sér. C11B, Vol. 16, Fol. 119. Estat des Batteaux, Gouelettes et Chaloupes, 1734.

1735 - AN, Col., Sér. C11B, Vol. 17, Fol. 90. Isle Royalle, 1735; and AN, Col., Sér. C11B, Vol. 17, Fol. 127. Etat des Batteaux et Chaloupes, 1735.

1736 - AN, Col., Sér. C11B, Vol. 18, Fols. 170-71v. Isle Royalle, Pesche et Commerce de l'année 1736.

1737 - AN, Col., Sér. C11B, Vol. 20, Fols. 21v-22v. Isle Royalle, Pesche et Commerce, 1738 (for 1737); and AN, Col., Sér. C11B, Vol. 20, Fol. 326. Etat général des Batteaux et Chaloupes, 1737.

1738 - AN, Col., Sér. C11B, Vol. 10, Fol. 220. Isle Royalle, 1738; and AN, Col., Sér. C11B, Vol. 26, Fols. 225-26. Isle Royalle, Pesche et Commerce, avril 1740; and AN, Col., Sér. C11B, Vol. 20, Fol. 334. Estat des Batteaux et Chaloupes, 24 décembre 1738.

1739 - AN, Col., Sér. C11B, Vol. 21, Fol. 152. Isle Royalle, 1739; and AN, Col., Sér. C11B, Vol. 26, Fols. 225-26. Isle Royalle, Pesche et Commerce, avril 1740.

1740 - AN, Col., Sér. C11B, Vol. 23, Fols. 160-60v. Isle Royalle, Pesche et Commerce, 1740.

1741 - France. Archives Nationales, Archives Marine (hereafter cited as AN, Marine), Sér. G, Vol. 53, p. 244. Le produit de la pêche de l'Isle Royale en 1741.

1742 - AN, Col., Sér. C11B, Vol. 26, Fols. 217-18v. Isle Royalle, Pesche et Commerce, 1743; and AN, Marine, Sér. G, Vol. 53, p. 248. Le produit de la pêche.

1743 - AN, Col., Sér. C11B, Vol. 26, Fols. 209-10v. Isle Royalle, Commerce et Pêche, janvier 1744; and AN, Col., Sér. C11B, Vol. 26, Fols. 217-18v. Isle Royalle, Pêche et Commerce, 1743.

1744 - AN, Col., Sér. C11B, Vol. 26, Fols. 227-28v. Isle Royalle, Pêche et Commerce, 1744.

1750 - AN, Col., Sér. C11B, Vol. 29, Fol. 206. Prevost au Ministre, 10 décembre 1750.

1751 - AN, Marine, Sér. G, Vol. 55, Fols. 274-75. D'après l'etat envoie par le Sr. Prevost.

1752 - AN, Col., Sér. C11B, Vol. 33, Fol. 437v. Prevost au Ministre, 24 décembre 1753; and AN, Col., Sér. C11B, Vol. 33, Fol. 496. Etat général des Batiments et Chaloupes en 1752; and AN, Marine, Sér. G, Vol. 55, p. 279.

1753 - AN, Col., Sér. C11B, Vol. 33, Fol. 436. Isle Royalle, 1753.

1754 - AN, Col., Sér. C11B, Vol. 34, Fols. 180v-81. Prevost au Ministre, 19 décembre 1754.

APPENDIX B. REGULATIONS CONCERNING THE EXPLOITATION OF THE COD FISHERY ON ISLE ROYALE, 20 JUNE 1743 [TRANSLATION]

(France. Archives Nationales,
Archives des Colonies,
F3, Vol. 50, pp. 254-59)

BY ORDER OF THE KING

Having caused the various practices which have arisen in the Isle Royale fishery to be reported to him, His Majesty has been informed that engagements of crews for the fishery are concluded under mere private signatures; that inhabitants sign such engagements indiscriminately at all times, even with fishermen who are in service elsewhere; that there are some who draw away fishermen and shoremen engaged with other inhabitants by granting them considerable increases in salary and, to compensate themselves therefore, sell their *engagés* the things they need at excessive prices, and often cause them to drink up or gamble away their salaries; that there are some also who refuse to settle accounts with their crews when asked by them to do so, and dismiss them without a *congé*, so that when these crews engage themselves with other inhabitants, this gives rise to many problems and disputes; that the accounts the inhabitants provide for their crews are usually not in order; that in the interval between the summer fishery and the fall fishery and during other periods the crews spend at the inhabitants' homes, they are fed there, but do nothing of any assistance to their masters; that there are even some who, during the course of the fishery, waste days doing nothing, others who do damage to vessels and fishing utensils, and still others who, in drunkenness or for other reasons, throw provisions overboard, make away with them, or sell them; that there are also crews who are unwilling to be engaged, and undertake to fish for themselves, without taking up residence or habitation in the colony; that there are travelling merchants who participate in the sedentary fishery for themselves, and return every year or two to France, without forming any establishment on the island; that there are ship captains who, having completed their fishery, leave part of their crews in the colony for the fall fishery, picking them up again the

next year and replacing them with others for the same purpose; that most of the inhabitants keep no books on the purchases they are obliged to make for their fishery; that there are some who, abusing their privilege of immunity from seizure of their habitations, vessels, and fishing utensils, inconsiderately take from ship captains far more things than they need for their fishery, without concerning themselves about whether they will be able to pay for them; that there are others who, when the fishery is over, use the proceeds therefrom to pay creditors who, in many cases, have only supplied them with superfluities, in preference to those who have provided them with the supplies necessary for their fishery, and that such maneuvres often turn out to be licenced for obligations contracted with such first creditors by inhabitants ostensibly for the provision of fishing supplies; and finally, that there are resident merchants in the colony who undertake to provide fishing proprietors with fishing supplies second-hand, which is a burden upon the fishermen and prejudicial to the trade of ship captains, over whom these merchants usually have preference when their supplies are to be paid for; and His Majesty, wishing to cause an end to be brought to these abuses, which are so contrary to good order, to the progress of the fishery and trade of Isle Royale, and to the establishment of the colony, has ordered and does order the following:

ARTICLE I.

In future, the inhabitants shall be required to sign the engagements of their fishermen and shoremen before a notary in places in the colony where there are notaries, or before a senior inhabitant appointed for the purpose by the *commissaire-ordonnateur* in places where there is none; the said inhabitants and crews shall promise by oath to give and receive only the salaries which are stated in the instruments of engagement, which shall include a statement that both parties have taken the oath. For the costs of the said instruments, only *5 sols* shall be charged for each man engaged, and these costs shall be defrayed by the inhabitants; and all other engagements which have not been signed in the form prescribed above shall be null, void, and of no effect.

II.

Inhabitants may not engage any fisherman or shoreman who does not hold a written *congé* from the master inhabitant with whom he served previously, nor any fisherman or shoreman newly arrived in the colony who does not hold a permit from the officer charged with the detail of classes on the island, or some other person appointed for the purpose; it is His Majesty's will that mention be made of such *congé* or permit in the engagements the inhabitants sign with their fishermen or shoremen, failing which the engagements shall be null and void.

III.

His Majesty most expressly forbids fishermen and shoremen to engage themselves with two different inhabitants for the same fishery, under penalty of full confiscation of their salaries; and the new master shall not be entitled to demand the return of anything he may have advanced to them; and one-half of what is confiscated shall go to the former master by way of compensation.

IV.

In order that the fishermen and shoremen not be drawn away during the course of the fishery, His Majesty has fixed the usual engagement time for April and October of each year, in the case of crews which are in the colony, this to apply to both the summer and the fall fisheries.

V.

The salaries of fishing crews shall be and remain fixed at 290 *quintaux* of cod for the crew of two chaloupes: 36 *quintaux* for each of the two masters, 32 *quintaux* for each of the other four fishermen, 40 *quintaux* for the shoremaster, 20 *quintaux* for the salter, 17 *quintaux* for the header, and 13 for the shore boy; and fishermen and shoremen shall not be entitled to demand anything further, upon any pretext whatsoever.

VI.

The oil and roe collected aboard chaloupes and prepared at the habitation shall go half to the inhabitant, and half to the crew; the crew's half shall be divided among the fishermen on the same basis as the cod each of them is to have under the terms of the previous article.

VII.

It is also His Majesty's will that shoremasters, masters of chaloupes, journeymen fishermen, salters, headers, and shore boys be liable for one-third of any damage to their chaloupes, tackle, and utensils in proportion to and up to an amount not exceeding what remains due them in salary, unless the said damage occurs through a grounding caused by bad weather, in which case they shall not be liable therefore.

VIII.

The salaries of crews fishing on vessels and schooners shall be reduced to three-sevenths of the catch. On the basis of 840 *quintaux* of catch to be divided among a crew of 11 persons, viz. one master or captain, one splitter, six fishermen, one salter, one header, and one boy or apprentice, the master or captain shall be paid 64 *quintaux* of cod, or eight shares out of three-sevenths of the catch; the splitter 36 *quintaux*, or four and one-half shares; the six fishermen 32 *quintaux*, or four shares each; the salter 32 *quintaux*, or four shares; the header 20 *quintaux*, or two and one-half shares; and the boy or apprentice 16 *quintaux*, or two shares; and the same proportions shall be observed with respect to crews made up of more or fewer than 11 persons.

IX.

It is His Majesty's will that the inhabitant receive the remaining four-sevenths of the catch taken by vessels and schooners, for the boat with its tackle and furniture; for the salt, fishing utensils, and provisions for the crew which he supplies; for his habitation, and for processing the cod on shore; for which work the inhabitant shall have shoremen in the number necessary and customary in the colony.

X.

It is also His Majesty's will that the said crews of vessels or schooners be liable for one-third of any damage to their boat, tackle, furniture, and fishing utensils during the course of the fishery, each in proportion to and up to an amount not exceeding what is due them in salary, unless the said damage occurs through a grounding caused by bad weather, in which case they shall not be liable therefore in any way.

XI.

Three-sevenths of the oil and roe collected and prepared aboard vessels and chaloupes shall be divided among the captain or master, the splitter, the fishermen, and the salter in proportion to their salaries; and the said crews shall not be entitled to demand anything further, upon any pretext whatsoever.

XII.

Shoremasters or salters who have salted the cod too heavily, using more than 10 *barriques* of salt per 100 *quintaux* of cod taken by chaloupes, shall be paid their salaries only in such over-salted cod; and further, the said shoremasters and salters shall be required to pay for the salt they have used over and above the said quantity.

XIII.

His Majesty abolishes the practice by which inhabitants grant passage or half-passage to their crews, both of chaloupes and of vessels and schooners and offer them other benefits directly or indirectly, expressly forbidding them to grant their crews anything more than what has been set forth above, under penalty of a fine of *50 livres* against offending inhabitants in respect of each fisherman or shoreman in whose favour they have violated this article, and a similar fine of *50 livres* against each fisherman or shoreman who had demanded and received from his master any benefit whatsoever beyond what is assigned him above; His Majesty including under the name of fishermen masters and captains; splitters, salters, headers, and boys making up the crews of vessels and schooners and going to sea therein; and half of the said fines shall go to the informer.

XIV.

For the fall fishery, which is carried on with chaloupes only, inhabitants may engage crews on a half-and-half basis, as has already been done by some inhabitants in the colony, viz. half the catch going to the fishermen and shoremen generally, who, out of their half, pay the master inhabitant for the provisions they consume during the fishery, and the other half going to the inhabitant for his chaloupes, tackle, fishing utensils, salt, and habitation; provided, however, that fishermen going out in chaloupes shall also be liable for one-third of any damage to their chaloupes, tackle, and fishing utensils during the course of the fishery, in proportion to and up to an amount not exceeding what is due them from this fishery, just as in the case of the summer fishery, except under the circumstances mentioned in Articles VII and X.

XV.

Each inhabitant may, in preference to others, continue to supply and debit the clothing and effects his crews may need, at the common retail price in the colony.

XVI.

Fishermen who remain ashore in the period between the summer fishery and the fall fishery while the shoremen finish making the cod and the cod are being weighed and delivered, and who are usually fed by the master inhabitant without doing any work, may not demand food from their masters during this period unless they work for the habitation, carting wood, mending the stages, repairing beaches and flakes, refitting vessels, schooners, and chaloupes, or doing other similar work of assistance to their masters.

XVII.

Fishermen shall likewise be obliged to work for the habitation in the spring while ice or bad weather prevents them from going out fishing.

XVIII.

It is His Majesty's will that the fishing days which fishermen waste through debauchery or laziness in the course of either the summer or fall fishery be deducted from them in proportion to what the other fishermen have fished on those same days.

XIX.

It is His Majesty's will that fishermen, shoremen, and other persons in the service of the inhabitants who have thrown provisions overboard, have wrongly made off with them, or have sold them, receive corporal punishment, according to the exigencies of the case.

XX.

His Majesty most expressly forbids inhabitants to advance to their crews, both fishermen and shoremen, more than two-thirds of what they have earned, except for clothing they need for the fishery; and if at the end of the summer or fall fishery it turns out, when the fishermen's and shoremen's accounts with the inhabitants are closed up, that the former are in debt to the latter for drink they have received from them in advance, it is His Majesty's will that the balance due the inhabitants be entirely lost to them; and the said fishermen and shoremen shall not be required to pay any part thereof.

XXI.

His Majesty enjoins each inhabitant to keep a marked and initialed account book, as pre-scribed by the ordinance on the subject of commerce, and to enter separately therein, day by day, the various supplies he has provided to his fishermen and shoremen, under penalty of losing the said supplies, in case of protest by the said fishermen and shoremen.

XXII.

The inhabitants shall be required to close up their accounts with their fishermen and shoremen in this book at the end of the fall fishery; and if a fisherman or shoreman asks him for a detailed copy of his account, the master inhabitant shall be obliged to provide him with it, particularly at the end of the fishery.

XXIII.

Nor may inhabitants refuse a written *congé* to their fishermen and shoremen who wish to engage themselves with someone else when the time of their engagements has ended; provided, however, that those who are in debt to their masters for clothing they have received from them shall themselves pay, or cause to be paid by the inhabitant with whom they will be going to serve, any debt they have with the master inhabitant they wish to leave; and to this end, master inhabitants shall be required to record such debt on the *congés* they issue to their fishermen and shoremen.

XXIV.

Merchants, passing travellers, and other persons not domiciled in the colony and not having their families their may not participate in the sedentary fishing upon any pretext whatsoever, under penalty of confiscation of their catches, their boats, and their fishing utensils; and one-third of what is confiscated shall go to the informer.

XXV.

His Majesty also forbids fishermen who are not domiciled in the colony and have no habitation there, and fishing crews being carried on ships which come to the island, to fish there on their own account, under penalty of full confiscation of their catches, their chaloupes, vessels, and schooners, and all their fishing utensils; and likewise, one-third of what is confiscated shall go to the informer.

XXVI.

Since fishing crews being carried on vessels which come to the colony might arrange with the captains of the vessels carrying them to fish for themselves, being provided with supplies by the captains, and since such captains might declare them, upon their arrival in the colony, to be fishing for the ships, mention be made thereof on the rolls of crew, and that captains be required to show the written engagements such crews have contracted with the shipowners.

XXVII.

His Majesty expressly enjoins the captains of ships coming to the colony who, at the time they leave the island to return to their destination, make a practice of leaving part of their crews ashore to participate in the fall fishery and spend the winter, to take all their people back aboard their ships, forbidding them to leave any member of their crews on shore, except in case of illness or other extraordinary circumstances; failing which they shall be subject to the penalties stated in His Majesty's ordinances on the subject of classes, and the catches taken by such crews left ashore in the colony shall be confiscated; and one-third of what is confiscated shall go to the informer.

XXVIII.

It is His Majesty's will that inhabitants, independently of the book they will be obliged to keep in future for the individual accounts of their crew members, also, indispensably, keep another book in the order prescribed by the ordinance, in which they shall, each day, enter debits and credits for the effects and goods they buy for their fishing and trade, and also the payments they make for such purchases, under penalty of being declared incompetent to fish in the colony.

XXIX.

It is His Majesty's will that the inhabitants pay for fishing supplies in preference to all other debts, such fishing supplies including chaloupes, vessels, or schooners for the fishery; tackle and furniture for these vessels; fishing utensils and salt; clothing and provisions for crews, and other things directly necessary for the fishery; and wine, brandy, tafia, and other drink they may buy shall not be deemed to be fishing supplies when the quantity thereof exceeds the settled quantity in the colony, which each inhabitant is obliged to give his crews as part of its provisions for each fishery.

XXX.

It is likewise His Majesty's will that in future, when inhabitants do not fulfill their engagements for fishing supplies, their moveable goods (if any), vessels, schooners, chaloupes, and fishing utensils be seized and sold to settle their debts; and if the proceeds from the sale of the same are insufficient, that their habitations be seized and sold through three consecutive calls for bids, a week apart, being awarded at the last call for bids, at the bar, to the last and highest bidder; provided, however, that vessels, schooners, chaloupes, and fishing utensils may not be seized for less than 30 *quintaux* of cod or 300 *livres* of indebtedness, nor habitations for less than 60 *quintaux* of cod or 600 *livres* of indebtedness incurred for purchases necessary for the fishery, to which effect His Majesty departs in this detail from his ordinance of 30 June 1723 and all other regulations contrary hereto.

XXXI.

His Majesty orders that ship captains who have, out of their cargoes, supplied inhabitants with effects directly necessary for their fishery be paid, in the year they provided such supplies, in preference to the inhabitant merchants residing in the colony and itinerant merchants who have provided the same effects to the inhabitants, when such captains and inhabitant and itinerant merchants have all provided supplies for the same fishery; His Majesty allowing to subsist, moreover the practice followed in the colony whereby the latest suppliers for the fishery receive payment in preference to suppliers for previous years.

XXXII.

And to prevent the wrong certain inhabitants might do to those who have provided them with fishing supplies by contracting obligations, ostensibly for fishing supplies, toward itinerant merchants and others who in reality have only sold them goods that are unnecessary for their fishery, it is His Majesty's will that such inhabitants and those toward whom they have contracted such obligations be prosecuted in the same way as forgers, and that they be punished to the full extent of the ordinances.

His Majesty orders and commands Sieurs Marquis de Beauharnois, Governor-Lieutenant General, and Hocquart, Intendant of New France, and Sieurs du Quesnel, his Commander, and Bigot, Commissary of the Marine and *Ordonnateur* on Isle Royale, to see to the execution of this regulation, which shall be registered at the Superior Council of Louisbourg, read, published, and displayed in all places where this is necessary. Done at Versailles, on the twentieth day of June, one thousand seven hundred and forty-three. Signed, LOUIS. And beneath, PHELYPEAUX.

For the King

Collated with the original by us, *Ecuyer*, Counsellor and Secretary of Finance to the King, House and Crown of France.

PARIS
IMPRIMERIE ROYALE
MDCCXLIII

(France. Archives Nationales, Archives
des Colonies, Sér. C11B, Vol. 14, Fols. 66–66v)

Statement of what it costs to equip two chaloupes for the fishery, supposing that the two chaloupes will take a catch of 600 *quintaux*

For the two chaloupes	40 q.	
For the sails	8 1/3	
2 grapnels @ 2 q.	4	
2 cables @ 4 q.	8	
2 pairs washboards @ 1 1/2 q.	3	
One *quintal* lines	8	
One *quintal* lead for sinkers	3	
150 hooks @ 2 q. per hundred	3	
4 [pairs?] oars @ 1 q.	4	
Cordage for mast rigging, buckets, salt shovels, and sail twine	2	
Pitch, tar, oakum, and nails	4	
	──	87 1/3 q.
25 *quintaux* biscuit @ 1 q.	25	
6 quarts peas @ 1 q.	6	
One *barrique* molasses	4	
1 1/2 *quintaux* butter @ 4 q.	6	
	──	41
One quart brandy	4	
80 *livres* salt butter @ 1 q.	80	
6 herring nets @ 3 1/2 q.	21	
2 mackerel nets @ 6 q.	12	
Rental of fish lot, stages, and huts	20	
	──	137
Wages		
Shoremaster	45	
2 chaloupe masters @ 38 q.	76	
4 companion fishermen @ 36 q.	144	
Salter	24	
2 boys @ 18 q.	36	
Passage of the said men @ 3 1/2 q.	35	360
	──	625 1/3 q.

(France. Archives Nationales, Archives des Colonies, Sér. C11B, Vol. 21, Fols. 297-304v)

Summer chaloupe fishery on the basis on which it is now conducted:

Salaries of crews for two chaloupes on the basis of a full catch of 600 *quintaux* between them:

2 chaloupe masters @ 38 *q.*	76 *q.*	
4 other fishermen @ 36 *q.*	144	
	220	220 q.

Shore crew to split, cut, and dry the cod and to fish for bait:

1 shoremaster	45	
1 salter	24	
1 header	20	
1 boy	18	
4	107	107

Passage of the 2 chaloupes masters, 4 other fishermen, shoremaster and salter when they come from France, (@ 4 *q.*
(When they have spent the winter in the colony, only half passage (2 *q.*) is paid. The crews also receive half the oil.)

	32	
	32	32

Provisions in common years for the subsistence of the 10 persons aforesaid during the summer fishery:

30 *quintaux* biscuit @ 1 1/3 *q.*	40	
2 *quintaux* butter @ 4 *q.*	8	
1/2 *barrique* molasses	3	
1 *barrique* guildive	8	
2 *barriques* peas @ 4 *q.*	8	
1 quart salt back fat	5	
	72	72

Effects and utensils for the fishery of
   the two chaloupes:
Depreciation for one season on 2 chaloupes
   costing 15 q. apiece new, and hardly good
   for more than 4 seasons   7 1/2 q.
Washboards for 2 chaloupes   2
Oars   3
Sails, good for only 1 year, @ 7 q.
   per chaloupe   14
2 mooring ropes @ 4 q.   8
2 grapnels @ 1 1/2 q.   3
50 fishing lines for the 2 chaloupes   9
cod hooks for *idem*   2
100 [*livres?*] lead for sinkers   4
8 herring and mackerel nets, for bait, @ 4 q.   32
[Depreciation] for 1 year on one-half
   chaloupe for setting and raising nets,
   costing 10 q. new, and good for 4 years   2 1/2
70 *barriques* salt, for the catch taken
   by the chaloupes, @ 1 q.   70
                                                      ——————
            Total   157   157
                                                                     ——————
                                                                     588 q.

Balance, full catch from two chaloupes   600 q.

Salaries, provisions for crews, and
   purchase of effects and utensils aforesaid   588
                                                      ——————
                                                      12 q.

   There remains for the fishing proprietor, out of the catch from two chaloupes, 12 *quintaux* of cod for his work; his habitation where there has to be a scaffold, huts for the crews, a fish lot, flakes, and fish lot utensils; the depreciation on any effects he has bought; damage that occurs in the course of the fishery; and the subsistence and maintenance of his family; out of this remainder of 12 *quintaux* of cod, the fishing proprietor must further pay any rebates on the cod.
   The same applies to the vessel and schooner fishery on the banks off Isle Royale.

APPENDIX E. EXTRACT FROM *MÉMOIRE AU SUJET DE L'ISLE ROYALE N.D.* [TRANSLATION]

(France. Archives Nationales, Archives des Colonies, Archives de la France d'Outre–Mer, Dépôt des Fortifications des Colonies de l'Amérique septentrionale, Ordre No. 141)

Balanced statement of the proceeds from fishing with two chaloupes, and the expenditure necessary for equipping of the same, for provisions, and for the salaries of 10 men employed therein during the course of the summer fishery, supposing that they each bring in 300 *quintaux* of cod.

Proceeds from the catch of two chaloupes   600 *q.*

*in the margin:
  Number of crew members and shoremen required:
    2 chaloupe masters
    4 companion fishermen
    1 salter
    1 shoremaster
    1 header
    1 shore boy
    —
  10 men

Expenditures for various fitting out and maintenance supplies:

2 chaloupes @ 5 *q.*	10 *q.*	
100 *aunes* of cloth for sails	10	
2 hawsers for mooring ropes @ 4 *q.*	8	
2 grapnels @ 1 1/2 *q.*	3	
1 cable for mooring the chaloupes at the stage	5	
2 pairs of washboards	3	
3 [pairs?] oars	3	
Cordage for mast rigging, moorings, and sail twine	2	
Pitch, tar, and nails for maintenance of the chaloupes	4	
Mess-tubs, pails, and compasses	1	
Nails for maintenance of stage	2	
1 *quintal* of lines	10	
1 *quintal* of lead bars	4	
200 hooks @ 1 1/2 *q.* per hundred	3	
4 herring nets @ 3 1/2 *q.*	14	
2 mackerel nets @ 5 1/2 *q.*	11	
80 *barriques* salt	80	173 *q.*

For purchases of provisions:

25 *quintaux* of bread, 2 1/2 *quintaux* per man, @ 1 q.	25	
5 quarts peas @ 1 q.	5	
4 *quintaux* butter @ 4 q.	16	
1 quart back fat	6	
1 *barrique* molasses	4	
2 *barriques* guildive	16	72

For salaries of 10 men and passage, as is customary:

2 chaloupe masters @ 38 q. on a basis of 600 q.	76	
4 companion fishermen @ 36 q.	144	
Salter	24	
Shoremaster	45	
Header and shore boy	36	
Passage for eight men @ 4 q.	32	357

Total expenditure	602 q.

Balance

Proceeds from the fishery	600 q.
Expenditure	602 q.
Deficit	2 q.

Consequently, the fishing proprietor is burdened with 2 *quintaux* of cod for the two chaloupes he sends out fishing, notwithstanding the fish lot, flakes, stages, and huts he provides, with all the utensils, at his own expense.

Note that the shares of the crews and shoremen increase or decrease according as the catch is greater or smaller, so that this changes nothing in the result.

(Québec, Archives du Séminaire de Québec,
Polygraphie 55, No. 44)

In the case of the cod which is fished by chaloupes during the summer, the chaloupes usually come in to shore every day to unload their catch onto the stage, whence it is first dumped onto a table. The man called the header, who holds a pointed knife with two cutting edges, guts the cod and breaks off the head; this is called heading. He then sends the cod across the table, to the splitter.   The splitter has a knife about 6 *pouces* long and about 16-18 *lignes* wide, with one cutting edge; it is thicker at the top, to make it heavier. The splitter opens up the cod and removes the bone, to about two-thirds of the way back from the head. He then drops it into a *traîneau* which is used by the salter to take the fish away to one side. The salter lays it down in rows, with the skin down, and covers it lightly with salt. He then piles the cod up, layer upon layer, repeating the operating with each layer. After the cod has been left in the salt for 3 or 4 days — sometimes twice this period of time, or more, if the weather is bad - it is put back into the sea in what is called a wash cage. It is washed, then brought back to shore to be placed in a pile, which is called stacking or stowing the cod. When the first fine weather comes, it is spread out, skin down, on platforms called flakes, or on the rocks on the shore. Then, before night, it is turned skin upwards. This is also done in case of rain. After it has dried somewhat, it is put into flackets of five or six fish, which are turned skin upwards for the night or in bad weather. The crew keep on spreading it out as described above for a greater or lesser number of days, depending on how fine the weather is, until it is partly dry. Then it is made into round piles in the shape of a dovecote. It is left there for a few days, and then is put back out in the air. It is turned, and made into larger piles of the same shape. It is left sometimes for a fortnight without being moved or spread out. Then, when it is almost dry, it is all left to sweat together. Once the weather permits, it is moved. This is called repiling. At the end, the cod turns out to be more or less fine, depending upon the weather and the season, how well the shoremaster knows his trade, and how vigilant he has been. Cod which is made in the spring, before the hottest weather, is usually the finest; and this grade is usually at a premium, particularly when the cod has been neither under- nor over-salted. Excessive salt causes it to remain whiter, and makes it crumbly; and it looks wet when the weather is humid. The cod which is caught in the fall — in October, November, December, and sometimes in January — remains in the salt until late March or early April. It is then washed and the same operations [as described above] are performed. It is not usually saltier than the other kind, but it is nonetheless not so highly prized. On vessels and schooners that stay out fishing for 20 or 40 days, the cod is headed and split on board, and when it is brought ashore is washed and cured in the same way as cod brought in by chaloupes, as described above.

ENDNOTES

Introduction

1 France. Archives de la Guerre, Archives du Comité Technique du Genie (hereafter cited as Archives Technique du Genie), Article 14, Pièce 48, Mémoire sur le commerce de l'Isle Royale par Mᵣ Prevost, ordonnateur, 1753.

2 See references in Charles De La Morandière, Histoire de la pêche française de la morue dans l'Amerique Septrionale (G.P. Maisonneuve et Larosse, Paris, 1962), 2 vols.; Harold Innis, The Cod Fisheries: The History of an International Economy (University of Toronto Press, Toronto, 1954), rev. ed.; Andrew Hill Clark, Acadia: The Geography of Early Nova Scotia to 1760 (University of Wisconsin Press, Madison, 1968); J.S. McLennan, Louisbourg from its Foundation to its Fall 1713-1758 (Fortress Press, Sydney, Nova Scotia, 1969); Mary C. MacDougall Maude, "The Settlements of Isle Royale and Isle St. Jean 1713-1758" (M.A. thesis, University of Toronto, 1965); and Christopher Moore, "Merchant Trade in Louisbourg, Isle Royale" (M.A. thesis, University of Ottawa, 1977).

3 Harold Innis, "Cape Breton and the French Regime," Royal Society of Canada Transactions, Section 11, 1935, pp. 51-87; and Andrew Hill Clark, "New England's Role in the Underdevelopment of Cape Breton Island during the French Régime, 1713-1758," The Canadian Geographer, Vol. 9 (1965), pp. 1-12.

4 David Alexander, "Newfoundland's Traditional Economy and Development to 1934," Acadiensis, Vol. 5, No. 2 (Spring, 1976), pp. 56-78. See also by the same author, "Development and Dependence in Newfoundland, 1880-1970," Acadiensis, Vol. 4, No. 1 (Autumn, 1974), pp. 3-31; and "The Collapse of the Saltfish Trade and Newfoundland's Integration into the North American Economy," Canadian Historical Association/La Société Historique du Canada, Historical Papers/-Communications historiques, 1976, pp. 229-48.

5 Anne E. Yentsch, "Understanding 17th and 18th Century Colonial Families: An Experiment in Historical Ethnography" (M.A. thesis, Brown University, 1975); "The Tenant Farmers of Naushon: A Report Based on Historical Documents and an Archaeological Report" (typescript Brown University, 1976); and "Marriage Patterns and Household Extension: 17th and 18th Century Cape Cod seen in Ethnographic Perspective" (typescript, Brown University, 1976). Although these papers have dealt primarily with tenant farmers rather than fishermen, the last paper held the promise of a future study correlating the kin-related coalitions on Cape Cod with subsistence fishing and agricultural patterns.

6 Christopher Moore, "Merchant Trade in Louisbourg, Isle Royale"; "The Maritime Economy of Isle Royale" in Canada: An Historical Magazine, Vol. 1, No. 4 (June, 1974), pp. 32-46; and "Commodity Imports of Isle Royale" (manuscript on file, Fortress of Louisbourg, 1975).

7 Nicholas Denys, The Description and Natural History of the Coasts of North America (Acadia), translated and edited by William F. Ganong (the Champlain Society, Toronto, 1908); Duhamel Du Monceau, Traité Général des Pesches et Histoire Des Poissons ... (Sailant & Nyon et Disant, Paris, 1769-77), 3 vols.

8 France. Archives Nationales, Archives des Colonies (hereafter cited as AN, Col.), C11C, Vol. 7, Fols. 3-11, Mémoire concernant la découverte, establissement et possession de l'ile de Terre-Neuve et l'origine des pecheries, 1710; [and] Reflections des negotians de Saint Jean de Luz et de Siboure quy representent a la cour en décembre 1716; ibid., C11B, Vol. 14, Fols. 62-66v, Répresentations des Habitants que font la pêche; ibid., Vol. 21, Fols. 297-304, Mémoire sur les habitants de l'Isle Royale [1739]; and Archives Technique du Genie, Art. 14, Pièce 48, Mémoire sur le commerce de l'Isle Royale par Mᵣ Prevost ordonnateur, 1753. See reference to additional such references in Christopher Moore "A Catalogue of Memoires" (manuscript on file, Fortress of Louisbourg, 1973).

Fishing and the Economy of Isle Royale

1   AN, Col., C11C, Vol. 8, Fols. 40-52, "Memoire sur les affaires presente du Canada et l'établissement du Cap Breton, 1706" by Raudot, Intendant of Canada.
2   See Moore, "A Catalogue of Memoires" for references to additional such memoires.
3   Christopher Moore, "Isle Royale in Canadian History" (paper presented for Professor J. Levitt, University of Ottawa, 1976), pp. 13, 21, and 28.
4   Moore, "Merchant Trade in Louisbourg, Isle Royale," p. 14.
5   Frederick J. Thorpe, "The Politics of French Public Construction in the Islands of the Gulf of St. Lawrence, 1695-1758" (M.A. thesis, University of Ottawa, 1973), p. 149.
6   Ibid., p. 14.
7   Moore, "Merchant Trade in Louisbourg, Isle Royale," pp. 30-34.
8   Ibid., p. 32.
9   Ibid., p. 33.
10 Ibid., p. 15.
11 Innis, "Cape Breton and the French Regime," Section 11, 1935, p. 64.
12 Moore, "The Maritime Economy of Isle Royale," p. 40.
13 James F. Shepherd and Gary M. Walton, Shipping, Maritime Trade and the Economical Development of Colonial North America (University Press, Cambridge, 1972), p. 51.
14 Moore, "Commodity Imports of Louisbourg," p. 28.
15 Innis, "Cape Breton and the French Regime," pp. 65-66.
16 McLennan, Louisbourg From its Foundation ..., p. 223.
17 Moore, "Merchant Trade in Louisbourg, Isle Royale," p. 25.
18 McLennan, Louisbourg from its Foundation ..., pp. 221-22.
19 Moore, "Commodity Imports of Louisbourg," pp. 16-19, 26.
20 Moore, "Merchant Trade in Louisbourg, Isle Royale," pp. 34-38.
21 Ibid., p. 37.
22 Clark, "New England's Role in the Underdevelopment...," p. 5.
23 Peter Bower, "Louisbourg: A Focus of Conflict 1745-1748" (manuscript on file, Fortress of Louisbourg, 1970), pp. 439-44.
24 Clark, "New England's Role in the Underdevelopment ...," pp. 5-6.
25 Ibid., pp. 3-4.

The French Fishery and the Establishment of Isle Royale

1   AN, Col., C11C, Vol. 7, Fols. 3-9, Memoire concernant la decouverte, les establissement et la possession de l'Ile de Terre Neuve et l'origine des pescheries, 1710.
2   Selma Barkham, "The Basques: Filling a Gap in our History between Jacques Cartier and Champlain," Canadian Geographical Journal, Vol. 96, No. 1 (February/March, 1978), pp. 8-19.
3   T.J.A. LeGoff, "Michel Daccarrette," Dictionary of Canadian Biography, Vol. 3, pp. 156-57.
4   Moore, "Merchant Trade in Louisbourg, Isle Royale," p. 43.
5   France. Archives Nationales, Archives de la France d'Outre-Mer (hereafter cited as AN, Outre-Mer), G3, Carton 2039-1, Pièce 66, Inventaire de la dame Veuve Dastarit, 13 février 1735.
6   See Innis, The Cod Fisheries ... for the most detailed account of this rivalry.
7   N.L. MacPherson, The Dried Codfish Industry (Department of Natural Resources, St. John's, 1935), pp. 16-17.
8   For more detailed accounts of fish drying see Atlantic Experimental Station for Fisheries, Halifax, "The Preparation of Dried Fish," Canadian Fisherman (December, 1925); D.L. Cooper, "Fish Drying," Progress Reports of the Atlantic Coast Stations, No. 20 (August, 1937); Ruth F. Grant, The Canadian Atlantic Fishery (Ryerson Press, Toronto, 1934), pp. 72-74; and MacPherson, The Dried Codfish Industry, pp. 16-43.
9   MacPherson, The Dried Codfish Industry, pp. 43-44.
10 Ernest Hess, "Studies on Salted Fish: 8. Effects of Various Salts on Preservation," Journal of the Fisheries Research Board of Canada, Vol. 6, No. 1 (1942), pp. 1-9; and MacPherson, The Dried Codfish Industry, pp. 24-25.
11 Ernest Hess and N.E. Gibbons, "Studies on Salted Fish: 10. Effect of Disinfectives and Preservatives on Red Haliophilic Bacteria," Journal of the Fisheries

Research Board of Canada, Vol. 6, No. 1 (1942), pp. 17-23..

12   H.P. Dussault, "Bacteriology of Light Salted Fish: Sliming," Progress Report of the Atlantic Coast Station, No. 55 (March 1953).

13   S.A. Beatty, "Putty Fish," Progress Reports of the Atlantic Coast Stations, No. 31 (February, 1942).

14   Innis, The Cod Fisheries ..., p. 50.

15   Ibid., pp. 48-49.

16   Gillian T. Cell, English Enterprise in Newfoundland 1577-1660 (University of Toronto Press, Toronto, 1969), p. 5.

17   De La Morandière, Histoire de la Pêche Française ..., Vol. 2, pp. 613-37.

18   C. Grant Head, Eighteenth Century Newfoundland: A Geographer's Perspective, Carleton Library Series (McClelland and Stewart, Toronto, 1976), pp. 11-13 and 15-16.

19   Robert Guitard, "Le déclin de la Compagnie de la Pêche sédentaire en Acadie de 1697 à 1702," La Société Acadienne: Les Cahiers, Vol. 9, No. 1 (March, 1978), pp. 5-21.

20   Head, Eighteenth Century Newfoundland ..., pp. 11-13.

21   AN, Col., B, Vol. 36, Fol. 47, Pontchartrain à Boyles et Jurats de St. Jean de Luz & Siboure, 1 février 1714.

22   Ibid., F3, Vol. 50, Fols. 150-50v, Ordonnance du Roi qui defend aux habitants non mariés à L'Isle Royale de Louer les graves et vignaux à ny appartinans a Mendou le 20 juin 1723.

23   Ibid., C11B, Vol. 5, Fols. 157-58, Ordonnance de Police Sur divers Cas du 27 avril 1720, St. Ovide et de Mezy.

24   Ibid., B, Vol. 45, Fols. 929-31, Ordonnance qui confirme celle rendue par les Srs. de St. Ovide et de Mezy le 15 7bre 1721. Le Roy, Paris, 12 May 1722.

25   Ibid., F3, Vol. 50, Fols. 254-59, Reglement concernant l'exploitation de la pêche de la Morue à l'Isle-royale. Le Roy, Versailles, 20 juin 1743.

26   Head, Eighteenth Century Newfoundland ..., p. 65.

27   Moore, "Merchant Trade in Louisbourg, Isle Royale," p. 25.

28   See Appendix A. References for Isle Royale Fishery Returns.

29   Great Britain. Public Record Office, Admiralty 1, Vol. 2652, Peter Warren, Boston, 9 July 1739, n.p. Conversion to livres based on average rate of exchange for 1739 given in John J. McCusker, Money and Exchange in Europe and America, 1600-1775: A Handbook (University of North Carolina Press, Williamsburg, 1978), p. 96.

30   AN, Col., F3, Vol. 50, Fol. 160, Sur les Representations que nous été faite par les maitres habitant pecheurs, et les marchands ..., St. Ovide et de Mezy, 21er 7bre 1724.

31   A.J.E. Lunn "Economic Development in New France 1713-1760" (Ph.D. thesis, McGill University, 1942), p. 477. Isle Royale values are taken from Table 4 in this work but have been halved to approximate colonial values.

32   Frederick J. Thorpe, "The Politics of French Public Construction in the Islands of the Gulf of St. Lawrence 1695-1758" (M.A. thesis, University of Ottawa, 1974), p. 2.

33   The Admiralty reports are found in AN, Col., C11B, Vol. 16, Fol. 119; and ibid., Vol. 17, Fol. 127; the civil authorities' reports are in ibid., Vol. 16, Fol. 257 and ibid., Vol. 17, Fol. 90.

34   See ibid., C11C, Vol. 9, Fols. 50-95; and ibid., C11B, Vol. 20, Fols. 21v-22v.

35   Innis, The Cod Fisheries ..., p. 161.

Land and the Isle Royale Fishery

1   Head, Eighteenth Century Newfoundland ..., pp. 182-87.

2   Innis, The Cod Fisheries ..., pp. 316-22.

3   Denys, The Description and Natural History....

4   For a compilation of legislation concerning the fishery at Isle Royale see Gilles Proulx, "Tribuneaux et Lois de Louisburg" (manuscript on file, Fortress of Louisbourg, 1975), pp. 35-41.

5   A.H. Clark, Three Centuries and the Island: A Historical Geography of Settlement and Agriculture in Prince Edward Island, Canada (University of Toronto Press, Toronto, 1959), pp. 29-30.

6   AN, Outre Mer, G1, Vol. 466, Pièce 81, pp. 112-13, Recensement par le Sr. De la Roque, Arpenteur du Roi, des Habitants de l'Ile Royale, commencé le 9 février 1752 et Ile St. Jean.

7   AN, Col., C11B, Vol. 15, Fols. 16v-17, Etat des Terrains actuelment occupés par divers habitant pescheurs Et autres particuliers ..., Louisbourg,

Isle Royale, 1734.

8   Ibid., Fols. 18-18v; AN, Outre Mer, G1, Vol. 466, Fol. 19, Pièce 83, Isle Royale, Louisbourg: Concessions, 1720-34.

9   AN, Outre Mer, G1, Vol. 466, Pièce 68, Recensement general des habitans Etablis a L'Isle Royale fait en l'année 1726.

10   AN, Col., C11G, Vol. 12, Fol. 59, Lettres patentes de concession de l'Isle St. Jean et de celles de Miscou en faveur de M. LeComte de St. Pierre, aoust 1719.

11   Ibid., Fols. 54v-56, Ordre du Roy au Sr. de la Boularderie pour faire la peche dans le Port d'Orleans, 15 février 1719.

12   AN, Outre Mer, G1, Vol. 466, Pièce 81, p. 52, Recensement par la Sr. De la Roque, Arpenteur du Roi ..., 1752.

13   Ibid., G3, Carton 2047, Part 1, No. 58, Bail à loyer de l'habitation, 5 mai 1743.

14   Ibid. G2, Vol. 209, Reel 2, Dossier 508, Plumitif pour servir à l'enregistrement des causes du Bailliage de Louisbourg, 11 octobre au 13 décembre 1754.

15   Ibid., Vol. 197, Dossier 143, Papiers concernant la succession de feu Elie Tesson La Floury, 1741.

16   Ibid., G3, Vol. 2041-2, No. 23, Vente d'un terrain, 6 avril 1752.

17   Ibid., G2, Vol. 183, Fol. 118, Affaires de Henry Nadau dit Lachapelle, 1734-36. I am indebted to Christopher Moore for bringing this and other references in this section to my attention.

18   Ibid., Vol. 204, Dossier 470, Fol. 39, Enregistrement de sentence, 21 juin 1752.

19   AN, Col., B, Vol. 59, Fols. 544v-45, Ministre à St. Ovide et Le Normant, 6 juin 1733.

20   Ibid., Vol. 61, Fol. 586v, Ministre à St. Ovide et Le Normant, 27 avril 1734.

21   AN, Outre Mer, G2, Vol. 180, Fols. 365-66, Reponse du Sr. LeBrun, 1726.

22   Ibid., Vol. 197, Dossier 143, Papiers Concernant la succession de feu Elie Tesson La Floury, 1741.

23   Du Monceau, Traité Général des Pesches, et Histoire des Poissons ..., Part II, Section 1, pp. 92-94; and Denys, The Description and Natural History ..., pp. 283-87.

24   See for example, Fortress of Louisbourg Archives, Map Collection, Map No. 1720-2, Plan de Louisbourg Avec Ses Augmentations faites pendant l'année 1720.

25   AN, Col., C11B, Vol. 33, Fol. 79, Ray-

26   mond au Ministre, septembre 1753.

26   Denys, The Description and Natural History ..., pp. 283-87; and Du Monceau, Traité Général des Pesches ..., pp. 92-94.

27   AN, Col., C11B, Vol. 15, Fols. 52-59.

28   Fortress of Louisbourg Archives, Map Collection, Map "Le Demy Bastion Dauphin avec sa Batterie qui defend le Port de Louisbourg par .17. canons de .24^L" ca. 1730

29   Denys, The Description and Natural History ..., pp. 287-88.

30   Fortress of Louisbourg Archives, Map Collection, Map No. 1734-4, Plan du Port et de la Ville de Louisbourg ... 1734. See Figure 8.

31   AN, Outre Mer, G3, Carton 2047-1, Pièce 151, Bail à loyer, 19 octobre 1750.

32   Denys, The Description and Natural History ..., p. 292; and Du Monceau, Traité Général des Pesches ..., Part II, Section 1, pp. 87-88.

33   AN, Outre Mer, G2, Carton 2057, Pièce 8, Inventaire des effets et habitation du Sieur Dessaudrais Robert, 1720.

34   Du Monceau, Traité Général des Pesches ..., p. 88

35   AN, Outre Mer, G2, Carton 2057, Pièce 8, Inventaire des effets et habitation du Sieur Dessaudrais Robert, 1720.

36   Denys, The Description and Natural History ..., p. 291.

37   De La Morandière, Histoire de la Pêche Française ..., Vol. 1, p. 169.

38   AN, Col., C11B, Vol. 15, Fols. 15-25, Etat des Terrains actuelement occupés par divers habitants pescheurs Et autres particuliers, Louisbourg, Isle Royale, 1734.

39   AN, Outre Mer, G1, Vol. 466, Pièce 81, p. 83, Recensement par la Sr. De la Roque, Arpenteur du Roi ... 1752.

40   Fortress of Louisbourg Archives, Map Collection, Map No. 1720-2, Plan de Louisbourg Avec Ses Augmentations faites pendant l'année 1720.

41   AN, Col., C11B, Vol. 15, Fol. 15, Etat des Terrains actuelement occupés par divers habitants pescheurs ..., Louisbourg, Isle Royale, 1734.

42   See Figure 6.

43   AN, Outre Mer, G3, Carton 2047-1, Pièce 151, Bail à loyer, 19 octobre 1730.

44   Ibid., G2, Vol. 204, Dossier 470, Fol. 39, Enregistrement de sentence, 21 juin 1752.

45   Prince Chitwood, "Excavation of the

Dastarit/DesRoches Property in the Dauphin Fauxbourg" (manuscript on file, Fortress of Louisbourg, 1977).

46 Denys, <u>The Description and Natural History ...</u>, p. 294.

47 Fortress of Louisbourg Archives, Map Collection, Map No. 1740, Plan du Fauxbourg de la Porte Dauphine a Louisbourg, Verrier, 1740.

48 AN, Outre Mer, G2, Vol. 183, Pièce 8, Papiers concernant La liquidation des droits de d.^{lle} Marguerite Guyon Veuve du Sr Bonnain Lachaume.

49 AN, Col., E, Vol. 93, Fols. 155-59, Inventaire chez M. de Costebelle, 7 octobre 1717.

50 Ibid., F3, Vol. 50, Fols. 254-59, Règlement concernant l'exploitation de la pêche à la morue à l'Isle Royale, Le Roi, Versailles, 26 juin 1743.

51 AN, Outre Mer, G2, Vol. 197, Dossier 143, Papiers concernant la succession de feu Elie Tesson La Floury, 1741.

52 Ibid., Vol. 181, Fols. 446-78, Papiers concernant la succession de feu Blondel, 1734.

53 Denys, <u>The Description and Natural History ...</u>, p. 280.

54 AN, Outre Mer, G2, Vol. 199, Dossier 188, Procès criminel du nommé Joseph Le Buf, 1744.

55 In 1739 Le Normant allowed 70 <u>barriques</u> of salt for a production of 600 <u>quintaux</u>. AN, Col., C11B, Vol. 21, Fol. 208, Mémoires sur les habitants de l'Isle Royale, 7 mars 1739.

56 AN, Outre Mer, G2, Vol. 195, Dossier 83 bis., Pièce 99, Compte avec C. Morin, 1733.

57 Ibid., Vol. 197, Dossier 132, Papiers concernant la succession de feu Elie Tesson La Floury, 1741.

Methodology of the Isle Royale Fishery

1 AN, Col., C11C, Vol. 9, Fol. 95, Estat General des Cargaisas des Batiments, Le Normant, 31 décembre 1737.

2 Ibid., Fols. 54-55.

3 Ibid., Vol. 8, Fol. 9, Reflections des negoçiants de Saint Jean de Luz et de Siboure quy represente a la cour en décembre 1716.

4 Ibid., C11B, Vol. 2, Fols. 60-65, Le Conseil de Marine, 20 avril 1717.

5 De La Morandière, <u>Histoire de la Pêche Française ...</u>, Vol. 2, p. 666.

6 See Appendix A. References for Isle Royale Fishery Returns.

7 Du Monceau, <u>Traité Général des Pesches ...</u>, Part II, Section 1, p. 92.

8 Archives Technique du Génie, Art. 14, Pièce 48, Mémoire Sur le commerce de l'Isle Royale par Mr Prévost, Ordonnateur de Laditte Isle, 1753.

9 Denys, <u>The Description and Natural History ...</u>, pp. 295-301; Du Monceau, <u>Traité Général des Pesches ...</u>, Part II, Section 1, pp. 91-94; and Figure 16.

10 AN, Outre Mer, G3, Carton 2047-1, Pièce 151, Contrat de construction, 24 septembre 1750.

11 Denys, <u>The Description and Natural History ...</u>, pp. 295-301.

12 Christopher Moore, "Street Life and Public Activities in Louisbourg: Four Studies for Animation" (manuscript on file, Fortress of Louisbourg, 1978).

13 AN, Outre Mer, G1, Vol. 466, Pièce 68, Recensement general des habitans Etablis a L'Isle Royale fait en l'année 1726.

14 Denys, <u>The Description and Natural History ...</u>, p. 273.

15 AN, Col., C11B, Vol. 1, Fols. 203-4, St. Ovide au Ministre, 2 décembre 1715.

16 Ibid., Vol. 14, Fols. 66-66v, Representations des habitants qui font la pêche, 1733.

17 Ibid., F3, Vol. 50, Fols. 254-59, Reglement concernalt l'exploitation de la pêche de la Morue à l'Isle Royale, 20 juin 1743.

18 Ibid., B, Vol. 59-2, Fols. 544v-45v, Maurepas à St. Ovide et Le Normant, 16 juin 1733.

19 Denys, <u>The Description and Natural History ...</u>, pp. 319-21.

20 Archives Technique du Génie, Art. 14, Pièce 48, Mémoire sur le commerce de l'Isle Royale ..., 1753.

21 Denys, <u>The Description and Natural History ...</u>, p. 322.

22 AN, Col. C11B, Vol. 21, Fols. 297-304v, Mémoire sur les habitants de l'Isle Royale, Le Normant, 1739.

23 See Appendix A. References for Isle Royale Fishery Returns.

24 AN, Col., C11B, Vol. 24, Fols. 41-42, Duquesnel et Bigot au Ministre, Louisbourg, 13 novembre 1742.

25 Ibid.

26 AN, Col., B, Vol. 59-2, Fols. 544v-45v,

Maurepas à St. Ovide et Le Normant, 16 juin 1733.

27 Archives Nationales, Archives de la Marine, C7, Dossier 221, Morpain Journal de Latour Cruchon dans le bateau Le diligent.

28 AN, Outre Mer, G2, Vol. 194, Pièce 63, Papiers concernant La Vente des hardes de deffunct Mathieu Laisné, Noel L'Eclanche, et Joannis L'Etapy Noyes au St Esprit, 1735.

29 Ibid., G3, Carton 2047-1, Pièce 35, Testament de Jacques Germain, 22 avril 1743.

30 Archives Technique du Génie, Art. 14, Pièce 48, Mémoire le Commerce de l'Isle Royale par Mr. Prevost, Ordonnateur, 1753.

31 AN, Col., B, Vol. 46, Fols. 172-79, Tres humbles remonstrances des negoçiants de St. Jean de Luz et Sibourne a Mgr. le Comte de Maurepas, 1724.

32 Québec. Archives du Seminaire du Québec, Polygraphie 55, No. 44, La secheries de la Morues se fait comme suit (1750s?).

33 AN, Col., B, Vol. 24, Fols. 303-4, Les officiers de l'Amirauté au Ministre, Louisbourg, 10 décembre 1742.

34 The account of fish dressing and curing is taken from Denys, The Description and Natural History ..., pp. 310-14, 331-38; Du Monceau, Traité Général des Pesches ..., Part II, Section 1, pp. 98-101; "L'Isle Royale en 1716," Revue d'Histoire de l'Amerique Française, Vol. 13, No. 3 (Decembre, 1959), p. 428; and Archives du Séminaire de Québec, Polygraphie 55, No. 44, La secheries de la Morues se fait comme suit (1750s?).

35 AN, Col., C11B, Vol. 8, Fol. 20, St. Ovide et Le Normant au Ministre, Louisbourg, 1 décembre 1726.

36 Ibid., Vol. 14, Fols. 110-13, St. Ovide au Ministre, Louisbourg, 20 octobre 1733.

37 Ibid., Vol. 18, Fol. 38, St. Ovide au Ministre, Louisbourg, 14 aoust 1735.

38 Ibid., Vol. 2, Fol. 86, M. Landrau par sa lettre du 3 avril 1717.

39 Du Monceau, Traité Général des Pesches ..., Part II, Section 1, pp. 97-98.

40 AN, Col., C11B, Vol. 17, Fol. 127, L'Officiers de l'Amirauté, Louisbourg 1735; ibid., Vol. 20, Fol. 136, L'Officiers de l'Amirauté, Louisbourg, 1737.

41 Ibid.

42 AN, Col., C7, Dossier 221, Morpain Journal de Latour Cruchon dans le bateau Le diligent.

43 Ibid., C11B, Vol. 21, Fols. 297-304v, Le Normant, Mémoire sur les habitants de l'Isle Royale, 7 mars 1739.

44 Ibid., B, Vol. 46, Fols. 172-79, Tres humbles remonstrances des negoçiants de St. Jean de Luz ..., 1724.

45 See Appendix A. References for Isle Royale Fishery Returns.

46 Ibid.

Participants in the Fishery

1 See Appendix A. References for Isle Royale Fishery Returns.

2 AN, Col., B, Vol. 45, Fols. 929-31, Ordonnance portant défense aux Capitaines et marchands forains d'engager les équipages des habitants, Paris, 12 mai 1722.

3 Ibid., C11B, Vol. 14, Fol. 232, Isle Royale, 1733; ibid., Vol. 16, Fol. 257, Isle Royalle, 1734; ibid., Vol. 17, Fol. 90, Isle Royalle, 1738.

4 Ibid., Vol. 14, Fol. 234, Liste generale des Batteaux, Gouelettes et Chaloupes, 1733; ibid., Vol. 16, Fol. 119, Estat des Batteaux, Gouelettes et Chaloupes, 1734; ibid., Vol. 17, Fol. 127, Estat des Batteaux et Chaloupes, 1733.

5 Clark, Acadia: The Geography of Early Nova Scotia to 1760, p. 307; Innis, "Cape Breton and the French Regime," pp. 59-61.

6 See Appendix A. References for Isle Royale Fishery Returns.

7 Clark, Acadia: The Geography of Early Nova Scotia to 1760, p. 306.

8 Assessments of the Isle Royale fishery determined the profitability of the shallop fishery but did not compare this with the schooner fishery. See for example, AN, Col., C11B, Vol. 21, Fols. 297-304v, Le Normant, Mémoire sur les habitants de l'Isle Royale, 7 mars 1739.

9 Archives Technique du Génie, Art. 14, Pièce 48, Mémoire sur le Commerce de l'Isle Royale par Mr. Prevost, Ordonnateur, 1753.

10 De La Morandière, Histoire de la pêche française ..., Vol. 2, p. 669.

11 Ibid.

12 France. Archives Départementales, Archives de la Charente-Maritime (La Rochelle), B, Amirauté de Louisbourg,

Reg. 272, Fol. 136, Déclaration d'arrivée, 28 avril 1743.

13 See Appendix A. References for Isle Royale Fishery Returns.

14 In the returns of the commissaire-ordonnateur, French fishing vessels are listed as being engaged in fishing and trading. See for example, AN, Col., C11B, Vol. 14, Fol. 232, Isle Royale, 1733.

15 See Table 12.

16 AN, Col., C11B, Vol. 14, Fols. 93-93v, Interdiction de mouiller des navires dans le havre de Niganiche du 15 août du 15 avril, St Ovide et de Mésy, 7 juin 1733.

17 AN, Outre Mer, G1, Vol. 466, Pièce 81, Recensement par le Sr. De la Roque, Arpenteur du Roi ..., 1752.

18 AN, Col., C11B, Vol. 1, Fols. 203-4, St. Ovide au Ministre, 2 décembre 1715.

19 Ibid., Vol. 21, Fols. 290-304v, Le Normant, Mémoire sur les habitants du l'Isle Royale, 7 mars 1739.

20 Ibid., Vol. 3, Fol. 119, Ordonnance obligeant à engager les pêcheurs et les graviers par contrat ecrit St. Ovide et de Mésy, Louisbourg, 30 aoust 1718; ibid., F3, Vol. 50, Fols. 83-84, Réglement de Police pour les habitant-pêcheurs et les marchands qui les fournessent, St. Ovide et de Mésy, Louisbourg, 1720.

21 Ibid., F3, Vol. 50, Fols. 254-59, Réglement concernant l'exploitation de la pêche de la Morue à l'Isle Royale, 20 juin 1743.

22 Ibid., C11B, Vol. 7, Fols. 7-7v, Ordonnance sur la préférence que doivent avoir les habitants pour acheter les morues et les huiles de leur compagnons-pecheurs, St. Ovide et de Mésy, Louisbourg, 21 septembre 1723.

23 Gilles Proulx, Aubergistes et Cabaretiers de Louisbourg, 1713-1758, Travail inédit No. 136 (Parks Canada, Ottawa, 1971), pp. 26-44.

24 AN, Col., F3, Vol. 50, Fols. 254-59, Reglement concernant l'exploitation de la pêche de la Morue à l'Isle Royale, 20 juin 1743.

25 AN, Outre Mer, G2, Vol. 198, Dossier 178, Papiers concernant le vente de trois Coffres de trois Compagnons Pecheurs du Service de françois Picard, 1743.

26 Ibid., G1, Vol. 466, Pièce 81, Recensement par le Sr. De la Roque, Arpenteur du Roi ..., 1752.

27 Ibid.

28 Ibid. The discrepancy between the data bases of the two tables comes from a fisherman whose age was given but not his place of origin.

29 AN, Col., C11B, Vol. 7, Fols. 68-74, de Mézy au Ministre, Louisbourg, 22 novembre 1724.

30 Ibid., Fol. 72, Rousseau de Souvigny, Louisbourg, 22 septembre 1724.

31 Ibid., B, Vol. 47, Fols. 276-91, Maurepas à St. Ovide, 26 juin 1724.

32 Ibid., C11B, Vol. 7, Fols. 68-74, de Mézy au Ministre, Louisbourg, 22 septembre 1724.

33 Ibid., Vol. 11, Fols. 42-45, Bourville au Ministre, Louisbourg, 14 décembre 1730.

34 Ibid., Vol. 12, Fols. 36-90, St. Ovide au Ministre, Louisbourg, 25 novembre 1731; ibid., Fols. 93-95v, Bourville au Ministre, Louisbourg, 28 novembre 1731.

35 Ibid., B, Vol. 59-2, Fols. 522-23v, Maurepas à St. Ovide et Le Normant, 19 mai 1733.

36 Ibid., C11B, Vol. 14, Fols. 43-50v, St. Ovide et Le Normant au Ministre, Louisbourg, 13 octobre 1733.

37 AN, Outre Mer, G2, Vol. 184, Fols. 518-44, Procédure criminelle contre Jacques Massé et Michel Pierre Bonneau, décembre 1737.

38 AN, Col., B, Vol. 59-2, Fols. 522-23v, Maurepas à St. Ovide et Le Normant, 19 mai 1733.

39 Ibid., C11B, Vol. 14, Fols, 43-50v, St. Ovide et Le Normant au Ministre, Louisbourg, 13 octobre 1733.

40 Ibid., Vol. 11, Fols. 42-45, Bourville au Ministre, Louisbourg, 14 décembre 1730.

41 AN, Outre Mer, G2, Vol. 184, Fols. 518-44, Procédure criminelle contre Jacques Massé et Michel Pierre Bonneau, décembre 1737.

42 AN, Col., C11B, Vol. 24, Fols. 111-19, Bigot au Ministre, Louisbourg, 4 octobre 1742.

Finances and the Fishery

1 Moore, "Merchant Trade in Louisbourg, Isle Royale," pp. 54-55.

2 AN, Outre Mer, G3, Carton 2047-1, Pièce 151, Contrat de construction, 24 septembre 1750.

3 Ibid., G1, Vol. 466, Pièce 81, pp. 111-12,

Recensement par le Sr. De la Roque ...,
1752.

4   Archives de la Charente-Maritime, B,
Amirauté de Louisbourg, Liasse 6121,
Pièce 106, Inventaire des habitations
des Isle De La Magdelaine, 5 novembre 1754.

5   AN, Outre Mer, G2, Vol. 190, Reg. 4,
Fol. 90, Estimation de terrain de pêche,
13 juillet 1733.

6   AN, Col., C11B, Vol. 14, Fols. 66-66v,
Representations des habitants qui font la
pêche, 1733; ibid., Vol. 21, Fols. 297-
304v, Mémoire sur les habitants de l'Isle
Royale, Le Normant, Versailles, 7 mars
1739; AN, Outre Mer, D.F.C., Amerique
Septrionale, Ordre No. 141, Mémoire au
sujet de l'Isle Royale, s.d.

7   Terence A. Crowley, "Government and
Interests: French Colonial Administration at Louisbourg, 1713-1758" (unpub-
lished Ph.D. dissertation, Duke University, 1975), pp. 202-3.

8   Ibid.

9   AN, Outre Mer, G2, Vol. 194, Dossier 63,
Papiers concernant la vente des hardes
de deffunct Matthieu Laisné ..., 1735.

10   AN, Col., C11B, Vol. 1, Fols. 203v-4,
St. Ovide au Ministre, 2 décembre 1715.

11   Head, Eighteenth Century Newfoundland: A Geographer's Perspective, p. 67.

12   AN, Col., C11B, Vol. 1, Fols. 203v-4,
St. Ovide au Ministre, 2 décembre 1715;
ibid., Vol. 14, Fols. 66-66v, Representations des habitants qui font la pêche,
1733; ibid., Vol. 21, Fols. 297-304v,
Mémoire sur les habitants de l'Isle
Royale, Le Normant, Versailles, 7 mars
1739; AN, Outre Mer, D.F.C., Amerique
Septrionale, Ordre No. 141, Mémoire au
sujet de l'Isle Royale, s.d.; AN, Col., F3,
Vol. 50, Fols. 254-59, Reglement concernant l'exploitation de la pêche ..., 26
juin 1743.

13   AN, Col., C11B, Vol. 1, Fols. 203v-4,
St. Ovide au Ministre, 2 décembre 1715.

14   AN, Outre Mer, G2, Vol. 194, Dossier 63,
Papiers concernant la vente des hardes
de deffunct Matthieu Laisné ..., 1735.

15   AN, Col., F3, Vol. 50, Fols. 129-29v,
Ordonnance interdisant aux marchands
forains d'engages les pêcheurs des habitants, St. Ovide et de Mésy, Louisbourg,
15 septembre 1721.

16   Ibid., C11B, Vol. 14, Fols. 90-91, Ordonnance fixant les avantages à fournir aux
pecheurs, St. Ovide et de Mésy, Louisbourg, 20 septembre 1733; ibid., Vol. 15,

17   Fol. 63, Ordonnance de police, St. Ovide
et de Mésy, Louisbourg, 22 mai 1734.

17   Ibid., F3, Vol. 50, Fols. 254-59, Reglement
concernant l'exploitation de la
pêche ..., 26 juin 1743.

18   AN, Outre Mer, G3, Carton 2041-1,
Pièce 4, Engagement avec Jean Milly,
1749.

19   Ibid., G2, Vol. 193, Reg. 1, Fol. 86,
17 aout 1752.

20   Ibid., Vol. 186, Fols. 450-53, Mémoire sur
les conditions générales faites aux
pêcheurs, s.d.

21   Ibid., Vol. 194, Dossier 63, Papiers concernant la vente des hardes de deffunct
Matthieu Laisne ..., 1735.

22   Same as Table 16 (see endnotes 7 and 8
this chapter).

23   See for example the rations specified in
Archives Marine, Sér. A1, Art. 57, Pièce
64, Vivres des vaisseaux, 11 septembre
1719.

24   AN, Col., C11B, Vol. 24, Fol. 322,
Ordonnance fixant la ration, Duquesnel
et Bigot, Louisbourg, 6 mai 1742.

25   Ibid., Fol. 323, Ordonnance pour
révoquer celle du 6 mai 1742, Duquesnel
et Bigot, Louisbourg, 23 septembre 1742.

26   Archives de la Charente-Maritime, B,
Amirauté de Louisbourg, Liasse 6113,
Pièce 57, Vivres d'hiver fourny a francois
gauthier, 1738.

27   AN, Col., C11B, Vol. 20, Fol. 418v,
Mémoire troupe, 1738.

28   AN, Outre Mer, G2, Vol. G2, Vol. 185,
Fols. 358-78, Succession de feu René
Herpin, 1739.

29   Archives Technique du Génie, Art. 14,
Pièce 48, Mémoire sur le Commerce de
l'Isle Royale par Mr. Prevost, Ordonnateur, 1753.

30   AN, Outre Mer, G2, Vol. 194, No. 53,
Fols. 223-26, Papiers concernant Marlan
Duhart, 1734.

31   Ibid., Vol. 181, Fols. 446-78, Papiers concernant la succession de feu Blondel,
1732.

32   Same as Table 17 (see endnotes 6 and 7
this chapter).

33   AN, Outre Mer, G2, Vol. 197, Dossier
143, Papiers concernant la succession de
feu Elie Tesson La Floury, 1741.

34   Same as Table 17 (see endnotes 6 and 7
this chapter).

35   AN, Col., F3, Vol. 50, Fol. 255v, Réglement concernant l'exploitation de la
pêche ..., 26 juin 1743.

36 Ibid., C11B, Vol. 14, Fols. 66-66v, Representations des habitants qui font la peche, 1733.

37 Moore, "Merchant trade in Louisbourg, Isle Royale," pp. 56-57.

38 AN, Col., F3, Vol. 50, Fol. 255v, Réglement concernant l'exploitation de la pêche ..., 26 juin 1743.

39 See Table 17.

40 AN, Col., C11B, Vol. 7, Fols. 11-11v, Ordonnance sur la préférence que doivent avoir les habitants pour acheter les morues ..., 8 septembre 1724.

41 Ibid., F3, Vol. 50, Fol. 160, Réglement fixant le prix de la morue entre ..., 21 septembre 1724.

42 Great Britain. Public Record Office, Admiralty 1, Vol. 2652, A State of the French Fishery at Cape Breton for June 1739, Peter Warren, Boston, 9 July 1739, n.p. Monetary conversion according to McCusker, Money and Exchange in Europe ..., p. 96.

43 Denys, The Description and Natural History ..., p. 305.

44 AN, Col., C11B, Vol. 21, Fols. 297-304v, Mémoire sur les habitants de l'Isle Royale, 7 mars 1739.

45 Compare prices for rum in AN, Outre Mer, G2, Vol. 198, Dossier 178, Pièces concernant la vente de trois coffres, de trois compagnon pecheurs du service de François Picard; ibid., Dossier 175, Pièces pour le compte de René Le Vasseur et ses gens, 1743.

46 Ibid., Vol. 183, Fols. 392-93, François Lessenne, Louisbourg, ca. 1733.

47 AN, Col., F3, Vol. 55, Fol. 255v, Réglement concernant l'exploitation de la pêche ..., 26 juin 1743.

48 AN, Outre Mer, G2, Vol. 197, Dossier 143, Papiers concernant la succession de feu Elie Tesson La Floury, 1741.

49 Ibid., Vol. 198, Dossier 178, Pièces concernant la vente de trois coffres ..., 1743.

50 Ibid., Vol. 194, Dossier 73, Jean Peré fils, 3 juin 1735.

GLOSSARY

Arimier. Stower, one of three fishermen in a shallop; later called a companion fisherman.

Batteau. Vessel; appears to have been used somewhat indiscriminately for any sailing vessel.

Beaussoin. Bosun; one of three fishermen in a shallop; later called a companion fisherman.

Brulee. Burnt; a dried cod burned by too much salt.

Chafaud (also échafaud). Fish stage; a wharf usually having a protected work area where inshore fishermen dressed and salted their catch.

Chaloupe. Shallop; a fishing boat of several tons burden, usually having a three-man crew.

Clai. A platform on which the washed salted cod were piled to drain before drying.

Compagnon-pecheurs. Companion fishermen; the two fishermen in a shallop under the shallop master's command, initially distinguished as the beaussoin and the arimier.

Decoleur. Header; the man in the dressing crew responsible for removing the heads of the cod; also refers to an apprentice position in the fishery.

Degrat. Fishing at a distance from the established base even if it was only a seasonal one; with schooners this was the grand degrat and with shallops the petit degrat; in the shallop fishery this involved making a temporary base away from the established one.

Echafaud. See Chafaud.

Falques. Washboards; collapsible canvas washboards attached to the gunwale of a shallop which were set up when the shallop was heavy laden or in large seas.

Galaire. See Clai.

Garçon. Boy, apprentice; usually refers to a fisherman or shoreworker acting in an apprentice or unskilled position.

Goelette. Schooner; a two-masted fore-and-aft rigged vessel.

Grand degrat. See Degrat.

Grave. Beach; usually of large gravel or small stones and used for drying cod.

Habilleur. See Trancheur.

Habitant-pecheur. Literally resident fisherman but more appropriately translated fishing proprietor.

Lavoir. Wash cage; a lattice-work cage set at the water's edge in which the salted fish were washed before drying.

Maitre. Master; the captain of a vessel.

Maitre de chaloupe. Shallop master; the fisherman in charge of the shallop's three-man crew.

Maitre de grave. Shoremaster; the man in charge of the dressing, salting, and drying of the catch in the shallop fishery.

Marchand. Merchantable; well-cured large cod.

Merluche. See Morue seche.

Morue seche. Dried cod; also called morue seiche.

Noues. Sounds; the air bladder along the backbone which was removed for food.

Petit degrat. See Degrat.

Picqueur. Throater; in a three-man dressing crew the throater cut the cod across the throat and down the abdomen before passing it to the header who removed the eviscera and broke off the heads; in a two-man dressing crew the header also performed the throater's duties.

Rances. Boughs or branches lying directly on the ground on which fish were laid to dry.

Saleur. Salter; the fisherman or shoreworker responsible for salting the catch.

Sapinette. Spruce beer; a popular brew among fishermen made from molasses and spruce boughs.

Timbre. See Lavoir.

Trancheur. Splitter; the man in the dressing crew who removed most of the backbone from the fish; also called an habilleur.

Vigneaux. Fish flakes; a platform made of wood for drying fish.